SOCIOLOGY OF AGRICULTURE

SOCIOLOGY OF AGRICULTURE

Technology, Labour, Development and Social Classes in an International Perspective

ALESSANDRO BONANNO

CONCEPT PUBLISHING COMPANY NEW DELHI-110059

ISBN 81-7022-281-2

First Published 1989

Alessandro **Bonanno** (*b.* 1954)

Published by
Ashok Kumar Mittal
Concept Publishing Company
A/15-16, Commercial Block, Mohan Garden
NEW DELHI-110059 (INDIA)

Lasertypeset by
Microtech Advance Printing Systems (Pvt.) Ltd.
H-13, Bali Nagar
NEW DELHI-110015

Printed by
Taj Press
DELHI-110006

Contents

Introduction

The papers contained in this volume were originally prepared for presentation at the XI World Congress of Sociology held in New Delhi, India, in August 1986. They also represent a sample of the ongoing research carried out by the members of the "Sociology of Agriculture" research committee, a group affiliated with the International Sociological Association.

The "Sociology of Agriculture" group was originally created in the 1970's as a response to profound changes which were taking place in agriculture. During that period it was evident that traditionally shared tenets in the area of agricultural studies were not adequate to explain an increasingly complex and unequal social setting. More specifically, the assumptions and predictions formulated by the proponents of the modernization school were shattered by a reality in which overproduction and famine, concentration of wealth and poverty, high use of chemical as well as industrial inputs in agriculture and the consequent destruction of environmental resources, and other similar contradictions went hand in hand. Accordingly, alternative research which could reformulate these and other topics was sought. More importantly, this research was sought in an international framework so as to stress the increasing interdependence of the various national settings and to acknowledge the intrinsic connection between development and underdevelopment.

After several years of informal work both within domestic contexts and at the international level, the "Sociology of Agriculture" group was officially constituted as a permanent organization of the International Sociological Association at the IX World Congress of Sociology held in Mexico City in 1982. On that occasion the basic research areas and theoretical emphasis of the group were reaffirmed. In particular, attention was paid to the definition of a common research agenda and to differences between sociology of agriculture and the parent discipline of rural sociology.

Since then, the central themes of research of the "Sociology of

Agriculture" group have gravitated around the contradictory development of agriculture and the effects its outcomes have on the various segments of the world population. Studies of group members have placed special emphasis on the scrutiny of the consequences of increased agricultural production and productivity. In this respect, the contradiction between the availability of agricultural surplus and the general phenomenon of overproduction were criticized in regard to the unequal distribution of food among social groups and countries. Furthermore, the inadequacy of theories explaining increases in the quantity of food produced as beneficial to producers and consumers alike and to the improvement of environmental conditions was underscored. The general position maintained by group members is that while increased food production may benefit large producers and agro-industrial corporations, it generates negative consequences for small and medium producers, for the economy at large, for the agriculture of developing countries, for the fiscal status of local states, and for the environment.

Similarly, sociology of agriculture has been concerned with the role of technology in agriculture and the patterns of technological development. In this context, the non-neutral character of technology has been emphasized together with the consequences that such a situation has generated. Accordingly, studies have been carried out stressing the labour-displacing dimension of the introduction of new technical devices and procedures, the cost of technological updating which discriminates against small and financially less endowed farmers, and the production and diffusion of size-biased technology.

The proletarianization of agricultural labour and the increasing socio-economic gap between North and South have been recurrent topics in the group's agenda. The crisis of petty commodity production, for example, has been explained as a consequence of increased concentration of resources in the hands of a limited number of producers, domestic and international policies which ultimately favour production of scale, market instability, and financial indebtedness of farmers. However, and in contrast to other analyses of similar topics, sociology of agriculture identifies these problems as socially generated, i.e. as the outcomes of specific actions undertaken by social groups. Accordingly, market instability as well as political-economic strategies are not seen as mere "accidents". Rather, they stem from wanted but also unwanted consequences of rational human actions.

This brief description of the investigating agenda of sociology of agriculture makes apparent the significant scientific departure from rural sociology. In a nutshell, sociology of agriculture and rural sociology differ at the substantive and epistemological levels.

Rural sociology has been traditionally concerned with rurality and has been understood as a discipline which studies phenomena as they occur in rural settings. In this respect, topics intrinsically related to the sociological tradition, such as crime and deviance, migration, family, religion, education, etc., have found their counterparts in rural sociological analyses. In essence, the empirical diversity of rural and non-rural settings has been crystallized at the scientific level with the development of discipline which studies rural phenomena as entities apart from the rest of society.

Sociology of agriculture is not directly interested in the study of rurality. Though rural events may take centre stage in some of its studies, its emphasis is on the totality of processes of production and distribution of food and their relationships to society. Accordingly, sociology of agriculture transcends the confines of rurality, though it does not exclude rurality from its agenda. Furthermore, it should not be considered a separate discipline from sociology. Rather, sociology of agriculture aims to be a substantive area within sociology whose boundaries are set by the specific characteristics of the issues investigated.

The difference between rural sociology and the sociology of agriculture extends to the epistemological sphere. Rural sociology has been traditionally dominated by an investigative posture inspired by logical positivism. Its enormous empirical production has been interpreted in a fashion in which reality is often equated with appearance and in which empirical is translated into empiricism. The recent development of sophisticated statistical techniques and the widespread use of computers have furthered the discipline in this direction. This is particularly the case in the United States, where studies outside the empiricist tradition are rather uncommon. To be sure, analyses carried out with alternative approaches have been part of rural sociology, yet they have never achieved the higher status they enjoy in other branches of sociology. Rural sociology has also traditionally been a functionalist discipline. Though it was born as an attempt to improve the standards of living of rural dwellers, it never projected itself outside the boundaries of existing socio-political conditions. Unlike the rest of sociology, rural sociology has intrinsically remained a politically conservative discipline despite some notable exceptions.

Sociology of agriculture stems from different epistemological roots. Its endeavours are based on a large and diverse background which includes neo-Marxism, neo-Weberianism, dialectic and critical theory, sociology of knowledge, and historical sociology, and which is characterized by an overall empirical orientation. Furthermore, and following the basic tenets of the above mentioned sociological schools, it maintains an understanding of "reality" in which action is ultimately evaluated against the content of socially shared fundamental values and ideologies. In this context, prominence is given to the scrutiny of social change and action in terms of the diverse interests existing in society. Accordingly, the achievement of progress is assessed, for example, in terms of the well-being of the less endowed classes, the satisfaction of popular needs, sustainability of life and the preservation of the environment rather than the accumulation of capital and profit generation. Sociology of agriculture, then, takes a politically progressive posture in which equality and freedom are to be shared by all segments of society rather than only a few and in which the analysis of the "is" must be always complemented with steps for the achievement of the "ought to be."

The volume is divided into three parts. In the first the relationship between the development and implementation of technology in agriculture is explored with respect to both new technologies, such as biotechnology, and conventional technology. The second part discusses the impact of changes in society and agriculture on the labour market, labour structure, and the relations of production in rural areas. The third and final portion of the book investigates patterns of rural development as they relate to agriculture, its structures, and its people.

The first part of the volume opens with a contribution by Lawrence Busch and William B. Lacy. In the paper the development of biotechnology and the potential consequences that its implementation and diffusion might have in the world food order are discussed. It is also indicated that biotechnology is a potentially revolutionary technology which refers to the development of biological products through the use of intact organisms or natural substances obtained from organisms. Accordingly, through biotechnology it is possible to replace certain aspects of agricultural production with chemically produced substitutes. Furthermore, it is possible to alter the characteristics of animals and plants to accommodate specific industrial needs. For example, the chemical creation of animal growth hormones can lead to their commercial availability with the results of larger and more productive species. Similarly, animal embryos can be modified through the introduction of new genes so that a permanent "new animal" is created with characteristics functional to higher levels of productivity. Also, agricultural commodities can be produce in vitro, i.e. in a laboratory rather than in the field.

Despite the obvious positive consequences that biotechnology can generate, the authors indicate the problematic repercussions that it can have at the spatial and international levels. It is pointed out that the possible shifting of agricultural production from the field to the lab can fundamentally alter the present city-hinterland relationship and replace it with one in which the hinterland becomes increasingly dependent on the city. Moreover, at the international level the possibility of creating substitutes for tropical crops may be devastating for the economies of Third World countries. In this respect, the possible elimination of markets for these commodities may very well signify the economic and social demise of these countries. Finally, the possible centralization of food production may be paralleled by a concentration of power in the hands of food-producing corporations.

Biotechnology is also the topic of study of the second paper contained in the volume. Shripad Deo and Navid Mohseni investigate the relationship between biotechnology and the development of agriculture in Third World countries. After a detailed analysis of the historical and theoretical aspects of the development of science and technology, the authors explore the historical conditions and patterns which have characterized the transfer of technology to the Third World. Furthermore, they assess the impact of new technologies on developing countries in terms of dependency and future outcomes. Their conclusions indicate the "nonrevolutionary" character of biotechnology. More specifically, they argue that the development and implementation of biotechnology does not address the needs of poor small farmers or the vast majority of dwellers in Third World countries. Indeed, the development of biotechnology fosters patterns of agricultural dependency already established by the green revolution. Increasing pauperization, concentration of power in the hands of developed countries, and the perpetuation of the existing structure are viewed as the major possible outcomes of the current development of biotechnology.

The overall role of technology and its relationship to agriculture are explored in two papers prepare by C.R. Prasad Rao and Susan Verghese. In the first of these "Land to Lab: Some Issues in Technology Creation in Modern Agriculture," the issue of the relationship between creation of agricultural technology and the satisfaction of the interests and needs of producers is addressed. In this context, it is maintained that there is an increasing gap between the actual needs, priorities, and conditions of use of technology and the assumptions under which technology itself is developed. According to the authors, the reason for this situation, typical of many developing countries such as India, is to be found in the lack of power on the part of producers in the technological decision-making arena. As a solution, a model of communication between farmers, extension agents, and scientists characterized by equal power among the parts is proposed. Furthermore, it is indicated that a larger role played by farmers in the process of creating technology can represent a step forward in the realization of a more harmonious relationship between man and nature. It can also signify an avoidance of the ecological and economic dangers associated with the development of technological packages imposed on farmers by food and chemical corporations.

The ecological question is explored in their second paper, "Development and Social Awareness: Perils of Pesticide Use in Modern Agriculture." In this contribution, the authors indicate that technology is commonly evaluated only within a discourse inspired by "instrumental rationality". Stressing an opposite viewpoint in which "substantive rationality" is foregrounded, they argue for more emphasis on cultural and value-oriented policies for the development of technology. With regard to the increased use of chemical pesticides in agriculture, they maintain that while the use of pesticides has had positive results, it has also created substantial ecological damage and has been recognized as dangerous to the health of human beings. They also point out that the use of chemical pesticides is the predominant form of pest control, as it is adopted to a much greater extent than alternative biological and integrated methods. The authors conclude that cultural values should occupy a central position in the creation and evaluation of technological innovations if degradation of the environment and harm to mankind are to be avoided.

The integration of cultural elements in the process of technological development is one of the issues discussed by M. Francis Abraham in the paper entitled "Appropriate Technology, Basic Needs, and Development: Where Are the Social Scientists?" Employing the case of India, the author analyses the socio-economic implications of the development and application of Appropriate Technology in rural settings. Appropriate Technology is an approach which emphasizes the use of a country's existing resources in the process of socio-economic development. More specifically, Appropriate Technology mandates the maximization of available resources, the minimization of those which are locally scarce, and emphasis on the region's cultural and historical characteristics in the development and implementation of technology. The author contends that many of the problems associated with the adoption of modern technology can be solved through substitution with Appropriate Technology. Consequently, efforts should be made to eliminate the existing obstacles to its implementation. In particular the gap between research and diffusion must be reduced as India has a rather developed research network and a far less developed delivery system. According to the author, of significant importance is the lack of involvement of social scientists, who are largely oblivious of the importance and potential impact of Appropriate Technology. Accordingly, its development and implementation are hampered as knowledge of sociocultural and psychological factors and cooperation between engineers, technologists, and social scientists are sporadic at best.

The second part of the volume opens with the paper entitled "Decentralization, Informalization and the Agricultural Labour Market in Advanced Western Societies." In the article I argue that the economic crisis of the past decade and the fiscal crisis of the State are among the principal factors which have reshaped the processes of accumulation and legitimation among countries of the First World. The crisis of "traditional forms of social production" has been partially overcome by decentralization of production (i.e. relocation of productive apparatuses in regions characterized by fewer labour disputes and a higher degree of State support) and by informalization of the production process (the reorganization of internal production in new and more flexible manners). The strategies of decentralization and informalization have been carried out in rural areas where their application is fueled by the availability of surplus labour in the agricultural sector. Indeed, the international farm crisis has frozen a segment of the agricultural labour force which is now willing to work in decentralized and informal productive operations. In turn, the situation has fostered the persistence of marginal farms, low levels of economic growth, and the further application of decentralization and informalization strategies.

The role of the State in agriculture is also explored in the paper, "Markets and Technical Change: The Case of the Portuguese Dairy Sector," which was prepared by Manuel Belo Moreira. In this case,

State intervention is viewed as a positive developmental factor in the modernization process of the Portuguese dairy sector. Implemented in a restored democratic setting after the collapse of the fascist dictatorship, State policies have stimulated milk production among small farmers. Through the development of cooperatives for the collection and commercialization of milk, small farmers succeeded in increasing not only productivity but also the quality of the product. Furthermore, the total number of operations increased, easing the severe unemployment and underemployment affecting the minifundio region of Portugal where milk production is concentrated. Though many problems still remain and further productive growth is necessary, the case of dairy production can be employed as an example of political economic action to be taken in a situation of underutilization of resources and a lack of capital for investments.

Employing the case of Venezuela, the remaining two papers of this section also discuss agricultural development and the role played by State sponsored policies. In the first of them, "The Capitalist Implantation by Substitution of Productive Forms in Venezuelan Agriculture" by Josefina Rios de Hernandez, induced capitalist development is discussed. It is argued that capitalist development in Venezuelan agriculture was not the result of an endogenous process of growth, but was largely stimulated by the use of "external" forces. This induced form of growth is best described by the concept of implantation, i.e. the preordained replacement of precapitalist forms of production with capitalist forms of production. The author confines her investigation to the process of implantation through substitution. In this case, implantation of capitalist agriculture is carried out by either private or State supported economic activities and is characterized by the mechanization of farms and changes in land tenure which alter traditional forms of organization of production and favour the development of capitalist relations.

The second paper, "The Capitalist Implantation by Colonization in Venezuelan Agriculture" by Nelson Prato Barbosa, analyzes "colonization", the other form in which implantation of capitalist relations of production occurred in Venezuelan agriculture. The author points out that this was the predominant form of change in agriculture and that it was largely financed through the revenues obtained by the international sale of oil, a primary export of Venezuela. He also underscores the different outcomes that the introduction of new technologies and the integration into a larger and internationally oriented market had on social actors in the rural context. Furthermore, and through the use of historical case studies, the characteristics of colonization by private initiative and colonization through State intervention are illustrated. The former refers to the process of settlement of new lands carried out by colonists, while the latter refers to State planning and sponsorship of colonization process in locations specifically selected for these purposes. It is concluded that in the case of both private and State sponsored colonization, the productive capacities of Venezuelan agriculture have been enhanced.

The third and final part of the book opens with an article by Vilma Figueiredo entitled "Small Farmers and Food Production in Southern Brazil: A Way Out of Imposed Technology." The thrust of the paper rests on the importance of popular participation in the developmental process. Through a historical overview, it is argued that the growth of the Brazilian economy in general, and of agriculture in particular, has been characterized by an "imposed path." Authoritarian regimes adopted strategies of economic expansion tailored to the interests of multinational corporations and international financial groups. Simultaneously, the interests of local farmers were not taken into consideration, resulting in the generation of numerous contradictions which ultimately hampered the creation of a more balanced and egalitarian process of economic growth. The establishment of a democratic regime in Brazil coincided with attempts to modify this situation. One example of this new course of action involves southern Brazilian farmers. Through efforts coordinated by local unions, agriculturalists have been able to create and implement an alternative agenda in which farmers' and people's needs are prioritized and in which production for the domestic market is of central importance. Though resistance to the new course is strong, some success has also been achieved. More specifically, farmers succeeded in being instrumental in the implementation the Law of Land Reform and in its success in the south. The author concludes that only through the participation of farmers in the political decision-making process can quantity and quality of food production be augmented.

Farmers' participation in the agricultural decision-making process is the central issue discussed in Ana Barbic's contribution, "The Development of Self-Management in Yugoslav Agriculture". Employing the case of the republic of Slovenia, the author reviews the events that led to the present situation of the Yugoslav agricultural sector. In particular an analysis of the role of agriculture in the post-World War

II reconstruction period is presented. It is maintained that agriculture occupied a secondary position in the adopted scheme of economic development and that an imposed process of cooperativization of farms was unable to generate adequate increases in agricultural production and productivity. Indeed, it was one of the fundamental factors responsible for the rural exodus which is still affecting Yugoslavia. Recently, legislation has partially corrected this situation, and hopes for the reversal of the rural exodus exist. However, the author continues, Yugoslav agriculture is characterized today by a socially owned (socialist) farm sector and a sector of privately owned farms. The former is largely inefficient and composed of farms too large to be effective from a productive point of view. The latter is characterized by farms which are too small to be economically viable. Moreover, inequalities both at the economic and social levels exist between agricultural and non-agricultural workers. A reinterpretation of the system of cooperation and an increased ability of farmers to participate in the decision-making process are viewed as possible alternatives to the status quo. Also, an increased reliance on market forces is considered to be a possible positive step toward the improvement of the socio-economic conditions of farmers and in the creation of a more efficient and environmentally sound agricultural sector.

The final paper of the anthology by K.S. Mandal is entitled "Equity Oriented Rural Development in India: Its Origin, Shaping and Consequences." In this work the author reviews the developmental programmes adopted in India since the 1960's. In the sixties the principal objective of the developmental policies chosen by the government was that of achieving a rapid economic growth through the promotion of intensive agriculture. In the following decade this orientation was changed in favour of a strategy based on equity. An equity-oriented posture also constituted the principal feature of developmental policies in the eighties. However, the author argues, equity-oriented rural development in India has been more a corrective measure than a true commitment to establish equality in the countryside. In fact, the outcomes of the developmental process have deviated from an equity-oriented course. It is concluded that the rural developmental strategy applied in India has predominantly benefited large producers, yet has reserved some benefits for small farmers. Though the latter have not been able to increase significantly their living standard, they have succeeded in guaranteeing their subsistence. In essence, equity-oriented rural development has not achieved substantial results nor has it simply been a symbolic measure. Rather, its review is mixed, as inequality has been furthered and the maintenance of minimum living standards for small farmers has been attained.

ALESSANDRO BONANNO

Contributors

M. Francis Abraham Professor of Sociology and Director of International Studies at Grambling State University of Louisiana, USA. His areas of interest are sociological theory and Third World development. He is the author of several books including *Modern Sociological Theory*, *Sociological Thought*, *Perspectives on Modernization and Women*, and *Development and Change*. He is a former student and member of the faculty at the Gandhigram Rural University of India and received his Ph.D. from Michigan State University where he had a Ford Foundation Fellowship. His research has focused on rural development in the Third World and the dynamics of modernization.

Ana Barbic Professor of Sociology and Political Science at the Edvar Kardeij University of Ljubljana, Yugoslavia. Dr. Barbic has carried out research and written on topics including family farming, structure of agriculture, farm self-management, role of women in farming, and socialist agriculture. She is the author of numerous papers which have been presented at national and international meetings and which have been published in several languages. Dr. Barbic is a member of the International Sociological Association, International Rural Sociological Society, Executive Council of the Research Committee on Sociology of Agriculture and vice-president of the European Society of Rural Sociology. She is the recipient of a Fulbright scholarship and a scholarship of the American Association of University Women.

Nelson Prato Barbosa Professor of Sociology at the Universidad Central de Venezuela, Caracas, Venezuela. He is also the editor of "Cuadernos del CENDES," an interdisciplinary scientific journal concerned with issues in the areas of social change and development. Dr. Barbosa has held teaching appointments at the Universidad de Los Andes and at the Universidad del Zulia, Venezuela. He is currently a member of the Socio-Historical Research Group at CENDES

(Centre for the Study of Development) of the Central University of Venezuela.

Lawrence Busch Professor of Sociology and Chair of the Committee for Agricultural Research Policy in the College of Agriculture at the University of Kentucky, USA. Dr. Busch received his doctorate in Development Sociology from Cornell University in 1974 and has been at the University of Kentucky since that time. He is the author and editor of numerous articles and several books on agricultural research, including *Science and Agricultural Development, Science Agriculture and the Policy of Research, Food Security in the United States,* and *The Agricultural Scientific Enterprise.* During the 1988-89 academic year Dr. Busch carried out research at the "Institute Francaise de Recherche Scientifique pour le Development en Cooperation" in Paris (France). Dr. Busch is the president of the Research Committee on Sociology of Agriculture of the International Sociological Association.

Shripad Deo completing his doctoral degree in sociology at the University of Kentucky, USA His scientific interests include agricultural biotechnology and its impact on agriculture and agricultural research, agricultural research in the Third World, technology transfer and the political economy of development and underdevelopment. He is the author of papers which have been presented at domestic and international professional meetings and which have been published in scientific journals and edited books. Among his most recent publications are: *Structure of Agricultural Research in the Third World.* In C.R. Caroll, J. Vandermeer and S. Rosset (editors) *Ecological and Agricultural Systems*, Macmillan, 1989 (co-authored with L. Swanson), and "Creating New Technological Dependencies," International Journal of Contemporary Sociology, Forthcoming (co-authored with J.de S. Silva, L. Busch, W. Lacy, N. Mohseni).

Vilma Figueiredo Professor of Sociology at the University of Brasilia, Brazil. A 1976 Ph.D. from George Washington University, USA, she has been a member of the faculty at the University of Brasilia since 1970. Dr. Figueiredo has also served as research fellow at the Centre de Recherches Internationales in Paris (France) from 1979 to 1982. She has authored numerous articles and books published in English, French and Portuguese. Presently she is directing research on State, Society and Agrarian Technology in Brazil. She is a member of the Board of Directors of the Brazilian Sociological Society and of the Social Sciences Research and Graduate Courses National Association.

Josefina Rios de Hernandez Professor of Sociology at the Faculty of Economic and Social Sciences of the Universidad Central de Venezuela, Caracas, Venezuela. She is a graduate of the University of Essex, England and a member of CENDES (Centre for the Study of Development) affiliated with the Central University of Venezuela. Her areas of interest include historical sociology and sociology of agriculture.

William B. Lacy Professor of Rural Sociology and Assistant Dean for Research at Penn State University, USA. He has been Professor of Sociology and Director of the Food, Environment, Agriculture, and Society in Transition (FEAST) programme at the University of Kentucky. He has co-authored and co-edited five books and numerous articles on agricultural research, biotechnology, and the structure of agriculture in the U.S., Sudan, India, and Brazil. He has been a consultant for the U.S. Congressional Office of Technology Assessment, the U.S. Agency for International development, and EMBRAPA (Brazil).

K.S. Mandal A Reader in Rural Sociology at the Gokhale Institute of Politics and Economics, Pune, India and a former member of the TATA Institute of Social Sciences, Bombay, India. He has researched and published on topics such as rural and agricultural development, social stratification, equality and poverty, and participated in professional conferences in India and abroad. Dr. Mandal is the author of the article "Influence of Dimension of Social Stratification on Benefit Distribution from Poverty Programmes," which was presented at the VII World Congress of Rural Sociology held in Bologna, Italy in 1988.

Navid Mohseni A doctoral candidate in the Department of Sociology at the University of Kentucky, USA. He is conducting research on "The Political Economy of the Computer Industry" and has written on topics such as technological transfer, international dependency, and technology creation.

Manuel Belo Moreira Professor of Sociology at the Istituto Superior de Agronomia of the University of Lisbon, Portugal. He has researched topics concerning the development of Portuguese agriculture, land reform and farm structure. He is also the author of numerous publications which have appeared domestically and internationally. Dr. Moreira is a graduate of the University of Grenoble (France) and a member of the Executive Committee of the Research Committee on Sociology of Agriculture of the International Sociological Association.

C.R. Prasad Rao Professor of Sociology at Andhra University, India, where he has been teaching and researching since 1967. His current scientific interests include sociology of agriculture, rural development and science, peasant studies, and technology and development. He has participated in numerous national and international seminars and conferences and has been professor and chair at the Reserve Bank of India's Council for Social Development, Southern Regional Office, Hyderadad, India.

Susan Verghese Lecturer in Sociology at Andhra University, India, where she has been a member of the faculty since 1973 and a doctoral faculty member since 1978. Dr. Verghese's research and teaching interests include sociology of education and social change.

PART ONE

AGRICULTURE
AND THE NEW TECHNOLOGICAL HORIZON :
BIOTECHNOLOGY AND DEVELOPMENT

1

BIOTECHNOLOGY AND THE RESTRUCTURING OF THE WORLD FOOD ORDER

LAWRENCE BUSCH & WILLIAM B. LACY

The past several years have witnessed the beginning of major change in world agriculture. For the first time in history, patents or patent-like protection has been accorded to plants and even microorganisms. The large agrochemical and pharmaceutical firms have bought many of the world's largest seed companies and have begun to integrate them into their larger plans. Computer technologies have begun to enter into the agricultural sector changing significantly the ways in which agriculture is managed. Finally, the new biotechnologies have opened new vistas in agriculture, making possible for the first time the engineering of improved plants and animals and promising to make the food processing industries more efficient than ever before. Together, these as yet unrealized changes are likely to dwarf the so-called Green Revolution of the 1970s.

In the USA farmers and the general public tend to view the new biotechnologies favourably. A recent article in American Vegetable Grower is indicative: "The promises of the new developments for agriculture have been widely publicized. For example, the potential and economic impact of new varieties of crops requiring little or no input of costly fertilizer is obvious. Similarly, the development of varieties with better resistance to disease and insects..." (Boyer, 1984:51). Unfortunately, our review of the literature and interviews with scientists suggest that the author's enthusiasm is at best likely to be short-lived.

In this paper we review these changes with specific emphasis on biotechnology. Of concern here is how the new biotechnologies and the social and scientific changes that they engender are likely to change the world food order. In the past, agriculture has been different from industry in part because of its need for land and its dependence upon seasonal changes. Lenin noted that "agriculture possesses certain features which cannot possibly be removed (if we leave aside the extremely problematical possibility of producing albumen and foods by artificial processes)" (1938:85). Yet, it appears that just such a possibility is upon the horizon. Is it possible that biotechnology will create a world of superabundant food? Is Hunger to be finally eliminated as a part of the human condition? Or are there other forces that will shape the biotechnology revolution to other ends ?

Saving the metaphor

Molecular biology arose in a large part due to the significant financial support that it received during the second and third decades of this century from the Rockefeller Foundation. The Foundation was interested in improving the theoretical underpinnings of medical research and thereby improving medical practice as well. Since human disease organisms are single-celled, the field of molecular biology focused on the molecular aspects of single-celled organisms. Moreover, since it was particularly easy to culture and manipulate, the common intestinal bacterium, E. Coli, became the research organism of choice among molecular biologists. Today, the genome of E. Coli is nearly completely mapped, and its structure and functioning are relatively well understood. The animal sciences and the food sciences benefited from this molecular move after World War II. The results of medical research were often transferable directly to meat and dairy animals while research into the nutritional basis of foods could often be used to develop improved food products and processing techniques. In contrast, plant biology has, until recently, rarely descended below the level of the plant organ or part. This was in part a result of lower levels of funding, but was also due to the greater difficulties of working with plant cells and the strong applied character of the plant sciences. Conventional plant breeding–what we may refer to as the old biotechnologies–has proceeded with some help from Mendelian genetics, but with a heavy and necessary dose of empiricism[1]. It has five basic steps:

(1) Discover or create genetically stable variation for the desired traits;
(2) Select those individuals best expressing those traits;

(3) Incorporate the traits into a desirable agronomic background;
(4) Test over a wide range of habitats and several seasons; and
(5) Release the new variety.

In the last decade, however, it has become possible to apply techniques developed for molecular biology to plant improvement. The techniques employed can be divided into two broad classes: culturing and genetic transfer. Culturing involves regenerating plants from protoplasts (plant cells with the cell wall removed), single cells, or plant parts. The advantage of these culturing techniques over conventional approaches are:

(1) One can grow an enormous number of different cells, each a potential plant, in a small area.
(2) The process of sexual reproduction can be bypassed allowing access to genetic material inaccessible through propagation of gametes and eliminating the delay necessary for multi-generational change. And,
(3) The mass screening of plant material can be accomplished in a very short time. For example, a harmful substance can be added to the culture so that only resistant cells grow. A problem with these techniques is that what is expressed at the level of the single cell may not be expressed at the level of the whole plant. In addition, only those traits for which a cellular analogue exists can be selected with these techniques; important traits such as root strength, resistance to lodging and other whole plant characteristics are not amenable to this selection procedure.

In contrast to culture techniques, genetic transfer techniques operate at the single cell, subcellular, or molecular level. Transfer may be either general or specific. General transfer involves fusion of protoplasts or transfer of specific organelles. Specific transfer depends upon the recombinant DNA technologies. A specific piece of genetic information is transferred and expressed in the host. For example, the genetic material conferring herbicide tolerance might be transferred and expressed in a crop plant. This would permit spraying for weeds without damaging the crop. This technique offers precision and control that is simply unavailable through culture techniques. In principle, breeders could transfer just the genes that they desire. The major limitation on recombinant techniques is the lack of information on the location and function of the genes of higher plants. Moreover, the most agronomically interesting techniques are multigenic; it is not yet clear just how these traits are expressed.

Most of these difficulties went unrecognized by early proponents of plant molecular biology. They often perceived plants as the same as microorganisms. As Chaleff explains: "With recognition of the similarities between cultured plant cells and micro-organisms came the expectation that all the extraordinary feats of genetic experimentation accomplished with microbes would soon be realized with plants. But because of the many ways in which cultured plant cells are unlike microbes, these expectations thus far have not been well-fulfilled" (1983:679; see also Schaeffer and Sharpe, 1983). In short, early proponents of plant molecular biology took their own metaphor literally: They believed that the metaphorical equivalence "plant=microbe" was actually the case.

Despite this confusion, plant molecular biology has been a growth industry. Vast sums of money have been invested by various public authorities, by large petrochemical and pharmaceutical firms, and, especially in the United States where laws are particularly favourable, by many small venture capital firms. Moreover, since the venture capital firms could only survive by attracting large sums of capital, they soon became the greatest defenders of the metaphor. The "hype" found in their prospectuses succeeded not only in attracting capital; it also convinced many in the public sector of the validity of the metaphor.

Though no major breakthroughs have occurred, significant progress has been made in developing plants tolerant to glyphosate (a major herbicide). This is relatively easy to do, largely an empirical as opposed to theoretical endeavour, and of enormous commercial potential to the companies involved. On the other hand, many of the venture capital firms have gone bankrupt, while others have been bought by the petrochemical or pharmaceutical giants. Only a handful have been able to generate some saleable product and generate additional capital.

Our story does not end here, however. While whole plants are much more complicated than microorganisms and pose many problems for molecular biologists, the metaphor might yet be saved if the focus of research were to change. Instead of focusing upon the improvement of plants in the field, plant cells could be treated as if they were microorganisms. The techniques of growth and fermentation of bacteria, already wellknown to certain segments of the food processing industry (especially cheese and bread making and alcohol production), combined with the newer techniques of cell culture, could be used to transform the production of certain agricultural commodities into industrial processes. In principle, any commodity that is consumed in an undifferentiated or highly processed form could be produced in this manner. Similarly, though with greater difficulty, tissue culture techniques could be used to produce edible plant parts in vitro. In short, agricultural production in the field would be supplanted by cell and tissue culture factories. Lenin's (1938) dream of the complete elimination of agriculture and its replacement with continuous process factory production would at last be fulfilled!

Of course, the transformation we have just described is not likely to take place within the next year or even the next decade. However, the scientific foundations for just such a transition are now being laid. Consider the following:

1. Markets for certain tropical commodities have already been restructured. Sugar, once an extremely important tropical commodity, has already been hardhit by the development of corn sweeteners and sugar substitutes such as aspartame (nutrisweet). It is unlikely that the sugar market will recover from its current state of overproduction and depression. It has been estimated that 8-10 million people in the Third World have had their livelihood threatened as a result (van den Doel and Junne, 1986)

A similar change has occurred with respect to the production of cooking oils; corn oil–an early biotechnology product–has become a major ingredient in prepared foods in developed countries. In addition, food processing techniques have been modified so as to make it possible to substitute oils in processing depending upon relative market prices. It is no longer uncommon in the United States to see bread packages that note that one or more of various oils may be used in the manufacturing process. In economic terms, the cross-elasticities of cooking oils have increased.

One of the first crops to feel the effects of tissue culture research was oil palm. The multinational Unilever Corporation, owner of large oil palm plantations in Malaysia and elsewhere, realized in the 1960s the potential for tissue culture of oil palm (James, 1984). Unlike many other tropical commodities, little breeding had been done with oil palm; at the same time, there was great variation among trees. The actual research was conducted in Britain in greenhouses. Palm oil production was increased by 30 per cent using tissue culture techniques to clone high-yielding trees. As of 1984, 12,000 improved trees had been planted. In addition, "more uniform, shorter trees facilitate mechanization of harvesting, reducing labour costs" (van den Doel and Junne, 1986:85). Given that annual replanting requirements are about 30 million trees, there is likely to be a steady market for the clones.

Moreover, as palm oil producers now compete with producers of other oils (coconut, soy, olive, cottonseed, etc.), the increased production of palm oil has had the effect of reducing demand for these other oils. While estimates of labour displacement are virtually unobtainable, it is clear that it has been significant. As more countries adopt the higher yielding palm clones, it is also likely that the price will be depressed to near or below production costs.

Still another area of significant scientific advance is the production of what food technologists call "fabricated foods." These fabricated foods "differ from conventional foods in that their basic components–proteins, fats and carbohydrates–may be derived from many sources and combined, along with the necessary micronutrients, flavours, and colours, to form an attractive product" (Stanley, 1986:65). Such foods may not even be identifiable as to origin to the consumer and might not even contain significant amounts of agricultural products.

2. Moreover, work is currently under way to produce the flavour components of expensive fragrances, spices and flavouring agents through tissue culture. As Colin and Watts (1983:731-2) put it:

> Many of the compounds are obtained from plants that are not grown under large-scale or controlled cultivation. This instability, combined with climatic, harvesting, transport difficulties, and possible political problems in the country of origin, often leads to considerable fluctuations in the price of the compound. With each of these compounds there is interest in a more stable and easily controlled source.

As one proponent puts it, "the research and development effort required is well worth the effort to achieve the *in vitro* production of not only speciality biochemicals, but potentially, food, spices, and industrial commodities" (Staba, 1983:203). In addition, work is apparently in progress to produce coffee, cocoa, rubber, and tea in vitro (Clairmonte and Cavanaugh, 1986; Tsai and Kinsella, 1981; Staba, 1985; Heinstein, 1981). These flavour components would be identical biochemically to the compounds naturally found in these products; hence, they would not be artificial in the sense now currently understood, but would be true equivalents. Citrus pulp vesicles have also been produced *in vitro* raising the possibility of daily production of fresh orange juice (Rogoff and Rawlins, 1986).

3. Also of significance is the factory production of pharmaceuticals and industrial chemicals using tissue culture techniques (e.g. Breuling, Aflermann, and Reinhard, 1985; Yamada and Fujita, 1983; Yamada, 1984; Misawa, 1985; Rosevear and Lambe, 1985; Anderson, Phillipson, and Roberts, 1985; Dixon, 1984). Many of these are secondary metabolites of plants which are extremely expensive, often selling for more than $ 2,000 per kilogram. They are used in various processes to produce dyes, astringents, pharmaceuticals [over 25 per cent of prescriptions filled in the USA are of drugs derived from plants (Balandrin et al., 1985)], and other rare but essential industrial or consumer goods. Balandrin et al. link *in vitro* production with political instabilities:

> As the natural habitats for wild plants disappear and environmental and political instabilities make it difficult to acquire plant-derived chemicals, it may become necessary to develop alternative sources for important plant products. There has been considerable interest in plant cell culture as a potential alternative to traditional agriculture for the industrial production of secondary plant metabolites. This has given rise to considerable research in Japan, West Germany and Canada (1985:1158).

Misawa (1985) makes a similar case, while noting the similarity between tissue culture and microbial fermentations. He also notes that such *in vitro* techniques permit production "in any place or season" (1985:60). According to Misawa, four types of products are currently of interest: alkaloids for pharmaceuticals, steroids also for pharmaceuticals, terpenoids as anti-tumour compounds, and quinones as drugs against heart disease. Systems as large as 20,000 litres-large by scientific, but still small by industrial standards–have been built (Fowler, 1984). Production costs are still too high for the use of these techniques with foodstuffs.

This interest has given rise to tissue culture-based manufacturing plants in both Germany and Japan. Balandrin et al. (1985) estimate that such processes are economical when cultures produce one gram of the desired compound per liter of culture and the selling price is between $ 500 and $ 1,000 per Kilogram. Among the advantages for culture techniques are standard growth conditions, less complex organization than the entire plant, and ease of purifying the desired product (Anderson, Phillipson, and Roberts, 1985). Moreover, the heterogeneity of cultured plant cells makes possible selection and multiplication of those that are highest-yielding. In addition, it should be noted that culture techniques offer the potential for the creation of whole new products, either through the collection of newly discovered, naturally occurring compounds or through the manipulation of cells so as to make them produce whole new compounds.

Consider the claim made by Yamada and Fujita (1983:726) for a dye/astringent currently being produced in Japan:

> Compared with shikon requiring 2-3 years for harvest of the plant, cultured cells permit harvesting within about three weeks, thereby greatly shortening the production period. The content of the shikonin derivatives in the cultured cells was about 14 per cent which was extremely high compared with 1-2 per cent in the normal field grown shikon.

Shikonin is now selling for over $ 4,000 per kilogram. One ironic note is that the *in vitro* production may actually drive the price down too rapidly, given the fact that the market is small and easy to saturate (Curtin, 1983).

Though we cannot get any confirmation or disconfirmation, persistent rumours exist within the genetic engineering community that at least one U.S company has mounted a programme to produce tomato pulp *in vitro*. This would be used to produce an array of canned tomato products including sauce, paste, catsup and puree. Similarly, it is rumoured that a french company has already produced apple sauçe *in vitro* and is now attempting to scale up the process for industrial production.

4. As noted above, the final stage of biotechnology involves the replacement of agricultural with industrial processes. Here, too, Much initial work has been accomplished. Cotton fibres have been grown directly from cotton cells. Unlike those produced in the field, test-tube fibres can grow from both ends. Though it is unlikely that field production will be replaced in the near future, the potential is there (BOSTID, 1986).[2] Similarly, the Japan Salt and Tobacco Public

Corporation and Plant Science Limited of the United Kingdom are both investigating the production of tobacco biomass for cigarettes *in vitro* (Curtin, 1983). DNAP is attempting to produce several dozen herbs and spices *in vitro.* So far the processes have proved too expensive for commercial use.

5. Perhaps the most far-reaching proposal to replace agriculture is that found in a still unpublished paper by Rogoff and Rawlins (1986). They propose a system in which most fields are planted to biomass perennial crops. These are harvested as needed and turned into sugar using enzymes at the point of harvest. Then, the sugar solution is piped to production plants in metropolitan areas. Finally, food is produced through massively scaled-up tissue cultures using sugar solution as a medium and nutrient source for the plant material. As they put it, "in this scenario, edible products are defined as foods synthesized from separately manufactured food components, or as plant organs, plant parts or their derivatives, produced as such and not derived from a whole plant grown in soil" (1986:7). Under this system, processing would be a year-round activity, with only what was actually needed being produced on a given day. Canning and freezing would be largely eliminated as would most spoilage, since food would be produced and consumed in the same general vicinity. In fact, the same production facility could be shifted from the production of one commodity to another in response to changing demand.

They argue that such a system would have other added benefits in terms of reduced need for monocropping, lowered soil erosion, less use of agrochemicals, reduced energy inputs, and minimal transportation costs. They go on to note that the work necessary to realize this scenario is already underway but that it is currently uncoordinated. Finally, they make a number of suggestions as to how to implement a more coordinated strategy.

The major stumbling block to such a transformation of the world food system is the production costs involved (Zenk, 1978). Yet, Rogoff and Rawlins note that transport costs would decline so much that production costs could double without significantly altering the final price (1986). Moreover, it is unlikely that the switch to such a system would take place all at once. It takes little imagination to realize that the areas where costs are likely to become competitive with conventional production methods are those where (1) the product is a luxury good available only in limited quantities, or (2) the product is a heavy one that is produced far from its point of consumption. In both instances cell and tissue culture techniques offer significant advantages. On the other hand, there are forces that will inhibit the rapid development of *in vitro* production. In particular, it appears likely that large, multinational enterprises will resist the rapid demise of lucrative markets that they now dominate, In these instances, they will intervene by defensive patenting, purchase of potential biotechnological competitors, or through government regulation of the competition. Indeed, Byé and Mounier (1984) have suggested that this is precisely why the otherwise desirable *in vitro* production of proteins (largely for animal production) has not become viable.

Since Third World nations rarely have significant shares of multinational companies, products that they produce and that cannot be produced in the field in more developed countries, are likely to become targets of biotechnological research. In addition, it should be noted that Third World countries are much more vulnerable to these production shifts than are those of the First or Second Worlds.

From Instruments To production Techniques

Before examining the probable impacts that such changes might have on the world food order, it is important to clarify the central role that science and technology play here. Traditionally, sociologists, philosophers and historians of science have viewed science as a quest for new ideas. They have paid scant attention to the central importance that instruments play in modern science (cf. Idhe, 1979; Busch, 1984; Lauden, 1985). Yet, it is obvious to even the most casual observer of a scientific laboratory that instruments are central to the process of doing science. Moreover, like the rest of us, scientists always work under fiscal constraints. This is true of even the most well-supported of scientists. As a result, they are always seeking new ways to cut the costs of their work, so that they might accomplish more with the same quantity of money. In this respect, then, scientists are much like capitalists; though they seek "credit" (profits?) in the form of publications and citations, rather than money, they are inexorably driven to reduce the cost of scientific production as surely as the capitalist is driven to reduce the cost of the production of goods and services.

When this process is applied to plant molecular biology we can immediately see its import: Plant molecular biologists are interested in understanding the ways in which plant genetic material is structured.

Even if they have little or no interest in the commercial opportunities presented by improved seeds, they need to have rapid, efficient methods for moving from individual protoplasts or cells to tissue to whole plants in order to insure that what is created at the molecular or cellular level is expressed in the whole plant. It must be remembered that the various techniques for plant improvement described above are nested as illustrated in Figure 1. Recombinant techniques may be used to transfer a specific gene to a plant cell. That cell must then be cultured into a tissue. The tissue in turn must be differèntiated into a plantlet (excluding for the moment the possibility of *in vitro* production). The plant must be tried in the field to see if the trait incorporated at the molecular level is expressed in a field setting. Finally, the new cultivar must be multiplied and released. If any of the steps in this process is impossible or difficult, it will become a bottleneck to the whole process.

This creates an enormous incentive to improve culturing processes. A recent article appropriately entitled "Assembly Line Plants Take Root" provides an illustrative example:

> The biggest thing keeping costs up and therefore holding tissue culture propagation back is the amount of labour needed to run a tissue-culture lab.... Automation is one solution to the problem of high labour costs. ARS (Agricultural Research Service) scientists in California have found a way to automate their tissue-culture work and at the same time achieve at least two to four times the growth rate of traditional tissue-culture systems. Plant geneticists Brent Tisserat and chemist carl E. Vandercook use a computer to control the flow of liquid nutrients to plantlets (Comis and Wood, 1986:10).

In addition, Daly (1985) has noted that automated DNA synthesizers are already in use in both academia and industry. Similar efforts are underway elsewhere, so as to make more efficient the plant improvement process. In short, the very development of scientific techniques and equipment for use in the improvement of conventional agriculture makes possible the elimination of agriculture as we know it. The metaphor, plants=microbes, can be saved by restructuring the process of plant improvement so that the culturing process produces the final product in terms of food or fibre; the field as a location for food and fibre production ceases to exist. As one scientist put it. "We have to stop thinking of these things as plant cells, and start thinking of them as new microorganisms, with all the potential that implies" (quoted in Curtin, 1983:657).

The consequences for the third world

Market Restructuring

It is wellknown that many Third World countries are dependent upon one or two agricultural commodities for their continued viability. Let us examine the impact of biotechnological replacements for one major tropical commodity: rubber.

Chamala (1985) notes that 75 per cent of the world's rubber comes from Malaysia and Indonesia. In Malaysia, 500,000 smallholders are directly engaged in rubber production while three million persons–approximately one-quarter of the total population–are directly or indirectly dependent upon it. In both countries smallholder rubber producers are among the poorest. The worldwide estimate is that about 22 million people depend on the rubber industry. If *in vitro* production of rubber or the use of tissue culture to improve guayule (which grows in temperate climates) (Fisher, 1986) were to become a reality, 90-95 per cent of these people would find themselves without a livelihood (assuming the other 5-10 per cent would continue to grow rubber for domestic and specialized uses). Given that the unemployment rates in Malaysia and Indonesia are already high, it would be unlikely that more than 20 per cent of those displaced (about four million people) would final alternative empioyment. Some estimated 16 million people with few or no marketable skills would find themselves without work or reduced to bare subsistence farming. This is a conservative without work or reduced to bare subsistence farming. This is a conservative estimate. Similar effects can be expected for other tropical commodities.

The market restructurings likely to be caused by *in vitro* production and/or the collapse of existing markets are enormous. Not only is it likely that a significant number of farmers and farmworkers will find themselves with no products to sell; even many of the traditional agribusiness giants may discover that their huge investments are no longer profitable.

Moreover, while markets may be protected from product competition, there is little or no protection to be had from new technology (Vergopoulos, 1985). As *in vitro* substitutes are developed, they are likely to lead to the rapid demise of existing commodity markets, as processors restructure their production toward the new input. Without some international system of regulation and compensation, the effects on particular countries are likely to be just short of catastrophic.

Table 1.1 : **Markets for Selected Plant Products**

Compound	Use	Wholesale Price US $/kg	Estimated world Demand in Millions of US $
Vinblastine	Leukemia	5000	18-20*
Jasmine	Flavour, fragrance	5000	0.5*
Sapota	Chicle	5000	18-20*
Caranthus			
Lithospernum	Shikonin	4500	?
Digitalis	Heart disorders	3000	20-55*
Rose otto	Rose oil	2800	12
Ajmalicine	Circulatory probs.	1500	5.25
Codeine	Sedative	650	50*
Papaver	Codeine, etc.	650	50*
Pyrethrins	Insecticide	300	20*
Buchu	Buchu oil	220	153
Cinnamon	Flavour	195	3.2
Quinine	Malaria, flavour	100	5-10*
Ginger	Flavour	100	33
Spearmint	Flavour, fragrance	30	85-90
Cinchona	Quinine	?	20-55*
Coffee	Beverage	12	2210
Cocoa	Beverage, flavour	4	891
Tea	Beverage	2	2917
Rubber	Tires, etc.	1	3565

*US market only.
Sources : Kenney, Buttel, and Kloppenberg, 1984; Curtin, 1983; FAO, 1985; Colin and Watt, 1983.

The degree of impact can be estimated by looking at past instances of market collapse. For example, Mexico used to grow large quantities of barbasco, a plant from which steriods can be produced. Today, barbasco is no longer grown; the large pharmaceutical companies have developed a method of chemically producing it without need for the plant (Dembo, Dias, and Morehouse, 1985). In the more distant past is the demise of the indigo industry in India. In 1897, 574,000 hectares of indigo were grown. By 1920, with the successful development of a chemically-produced indigo, field production had all but ceased.

Though the planters attempted to develop a research fund so as to compete with chemically-produced indigo, they ultimately failed to do so (Martin-Leake, 1975). The result, for that region of India, was widespread depression and unemployment (Kenney, Buttel, and Kloppenburg, 1984).

Moreover, every time that farmers and farmworkers are displaced, they migrate to urban areas in search of work. Often men leave their wives and families behind to engage in subsistence farming. If the scenario developed here proves accurate, we can expect the same pattern on a massive scale. The impact could be particularly hard on women and children for they are likely to bear the full brunt of market restructuring.

It should also be remembered that market restructuring affects not only the producers of the primary product but also those engaged in the transport, processing, and even retailing of the final product. Thus, cumulative effects are likely to be much greater than would be the case if changes were somehow limited to farm production. Among the possible effects are geographic shifts in the location of the production process, the demise or growth of secondary industries, secondary effects as the former producing countries are unable to afford importation of various manufactured goods, and the decline of certain consumer goods industries in the Third World as demand declines.

Finally, it requires no crystal ball to realize that a very significant effect of *in vitro* production will be an increase in the already high Third World debt and a concommitant deficit in the balance of payments in Third World countries. While this will undoubtedly affect those nations, its effects are likely to be felt even more strongly in the West. The possibility of major bank failures is there even now and can only be exacerbated by market restructuring. Moreover, we can safely assume that the major banks will not be allowed to fail, but will be bailed out by the national banks of the Western powers. This, in turn, will tend to raise taxes and national debts in the West.

Third World Science

In recent years certain Third World nations–for example, India and Brazil–have succeeded in establishing significant agricultural scientific communities. These scientists have only recently begun to develop products that can make a substantial difference in their respective agricultural economies. They have been aided in this endeavour by the

IARCs and bilateral aid agreements that, in turn, have provided training and support for these new scientific communities.

Nevertheless, Third World science has often tended to be derivative in character (Chatelin, 1986). It has not entered into uncharted areas but has been content to take care of the details in the well-charted parts of nature (Goonatilake, 1984). It has avoided the so-called "basic sciences" on the grounds–usually supported by First World donors–that all basic science was an expensive luxury. As Daniel Goldstein has put it, "Original scientists in underdeveloped countries work in a social vacuum, and their activity is considered perhaps ornamental, but always useless" (1986:6). As Chatelin and Arvanitis (1984) have put it, the Third World has been characterized by "scientific domination."

As a result, even in those Third World nations that have scientific communities, few scientists are capable of pursuing the paths opened by the new biotechnologies. Only a few Third World nations have the critical mass of scientists necessary to engage in genetic engineering; a somewhat larger number may be able to employ tissue culture techniques for selected crops. None will be able to mount the broad research campaign needed even to stay abreast of the First World countries. While the UNIDO-supported international center for biotechnology might help in some small ways, it is unlikely that it will ever be able to take on the role played by the IARCs, in supporting the research that led to the Green Revolution.

The rapid worldwide growth of technical personnel in this area is also worthy of note. For example, in 1931 only two persons were working in the area of plant tissue culture; now there are more than 10,000 persons (Gautheret, 1983) nearly all of whom are located in the developed world. Salaries for such persons are high and training programmes are still relatively few. Only a handful of developing countries will be able to afford to create a critical mass of scientists in this area.

Access to final products as well as to the processes used to produce them is another concern. The new technologies generated by the Green Revolution were created in the public sector and access to them was relatively unproblematic. In contrast, much of the biorevolution is likely to be shrouded in secrecy. This is particularly true in the USA. As Curtin (1983:649) puts it: "companies tend to be more close-mouthed about their activities in plant tissue culture than they are about culturing mammalian cells, especially in the U.S. Elsewhere -- in

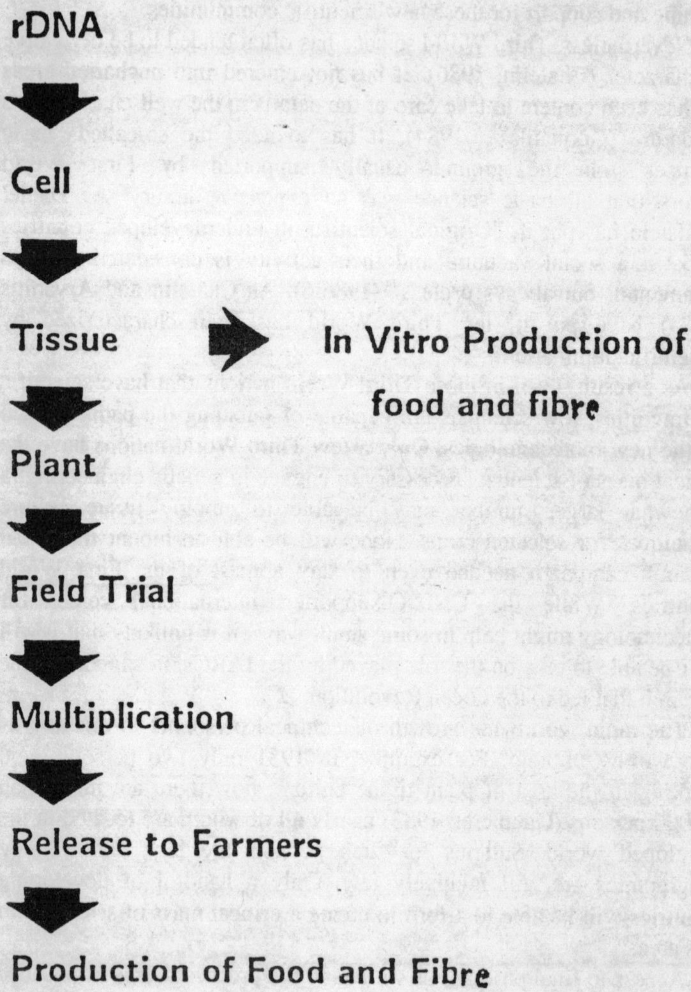

Fig. 1.1
(Steps Involved in Plant Improvement and Production Using Conventional Techniques and the New Biotechnologies)

Western Europe and, most notably in Japan – their interest is more substantial and more obvious." This secrecy means that the biorevolution is likely to appear with shocking immediacy as once stable markets collapse and market shares are suddenly rearranged.

Moreover, the Green Revolution was spatially limited by the very nature of the crops involved (Buttel and Barker, 1985). The biorevolution will not be spatially limited as the only limit on factory production of food is costs incurred as a result of the distance from the manufacturing plant to the market. Therefore, those products now consumed in the West are likely to be produced there as well.

In addition, even if the Third World nations are able to develop a cadre of trained scientists, they must also overcome the barriers currently restricting access to the instruments, supplies, and materials needed to do biotechnological research. Without this, at best they can copy some of the less imaginative efforts of the West. David Baltimore (1982:33) explains:

> However, that training and the ability to self-define problems were coupled to an infrastructure of supply companies that provided specialized equipment such as centrifuges; enzymes, which are the heart of the technology; fine chemicals; laboratory gadgetry, some of which is crucial to research; specialized chromatography media; and so on. The infrastructure is possibly the most crucial part of the system.

Moreover, no research group can manufacture all of the chemicals it needs – or even most of them – itself. It must depend upon commercial suppliers that are remote from the point at which the research is conducted. Many of these materials – especially the enzymes – are fragile, must be shipped under special conditions, and must be delivered within several days. For example, one large US supplier of enzymes, the Sigma - Alrich Corporation, was able to sell $ 215 million in chemicals last year, almost entirely by mail. The company reports that 97 per cent of their orders are shipped within 24 hours of receipt (Simon, 1986). Most Third World countries will have great difficulty in linking their scientists to such a supply system.

Let us assume, however, that the Third World is successful in mounting the expertise necessary to develop its own biotechnology research competence. This might be of some small help to offset the collapse of various commodity markets. In some cases patents developed in the Third World might even prevent developed countries from capturing certain markets. Tissue culture might be used to reduce production costs of field-grown crops thereby making competition from *in vitro* production that much more difficult. However, while this might slow down the changes, it would in no way prevent the loss of millions of jobs in agricultural production. In short, the development of Third World science is no solution to the destabilizing effects of biotechnology.

Political Stability

It should be apparent by now that the *in vitro* production of food is very likely to increase political instability. In the Third World, it is likely to dash the hopes of many who yearn for a better life. While some persons would undoubtedly revert to food crop production on land now used for cash crops, popular upheavals would be likely as displaced populations demanded that their fragile states provide food and shelter. In the First World it is also likely to create political instability, as a result of bank failures, higher taxes, and the elimination of certain processing and transport industries. In both the First and Third Worlds, it is likely that the changes forecast here will create additional social stratification, a greater gap between the rich and the poor, and greater possibilities for class conflict.

Nutrition and Food Safety

It also appears unlikely that *in vitro* production will be without effects on both nutrition and food safety. For those who are displaced by *in vitro* production, the spectre of hunger may loom ever larger. Somehow, there is a continuing failure among scientists to realize that the production of more food does not necessarily lead to the elimination of hunger. Hunger can only be eliminated when everyone has the wherewithal to obtain the means of subsistence, either through access to land on which to grow food or to money to purchase food.

For those in the Third World who do have enough to eat and for those in the First World who become consumers of food (and drugs) produced *in vitro*, the problem is somewhat different. We now know that human nutrition depends upon the daily ingestion of certain nutrients although some disagreement exists as the precise quantities. In general, we can say with some confidence that we know the quantities of the macronutrients that we need to consume but that we have somewhat greater difficulty with the micronutrients – in part because of the small quantities involved, but also because of the difficulty of conducting tests on human beings. We know even less about the interactions amongst foods or between foods and adulterants.

What all this adds up to is a very large question mark. Every time that we eat fresh foods, we ingest unknown quantities of unknown soil bacteria. Do these play a role in our diet? Every plant product that we eat is itself the result of the peculiar circumstances that happen to exist in a particular field. This in turn affects the flavour and even texture of the things we eat. Normally, we eat a variety of foods that receive their nutrition as a result of these different field conditions. *In vitro* production, on the other hand would unify these conditions and perhaps eliminate many of them. Would such a food supply be adequate, not only in the short run but also in the very long run? What effects might it have after several generations?

Unfortunately, the present state of science falls down in its attempts to answer these questions. Indeed, it appears that they are very difficult if not impossible to raise. Consider the following:

1. There is a sharp break between the food and nutrition area and that of agricultural production (see Randolph and Sachs, 1981). This gap extends from the US National Academy of Sciences where food and nutrition and agriculture are under separate boards, to the structure of most universities where colleges of agriculture are often separate from schools of nutrition.
2. Both the agricultural and nutrition sciences tend towards reductionism. That is to say, they tend to assume that the world can be subdivided into a series of discrete problems that can be resolved serially. Without a doubt, for some problems, this type of approach can and does work quite well. However, for other problems such as those raised here, reductionist strategies fail to grasp the problem.
3. Even now the testing of new food products and food additives is a complex and arduous task with which few governments are able to cope adequately. The production of *in vitro* food and drug products will undoubtedly put added stress on a system that is at best a weak one. Will these products be considered "the same" as food produced in the field and so escape government testing? Or will they be considered "new" and

require elaborate test methods? (We should note here in passing that they are likely to be considered "new" from the standpoint of obtaining patents). If they are to be tested, how stringent should these tests be? Indeed, can tests be devised for problems that are chronic rather than acute?

4. The emphasis in both nutrition and food science is not upon the development of informed policies, but on the clinical diagnosis of problems in the former and the development of new food products and processes in the latter. Members of both sciences often studiously avoid the food policy questions on the grounds that they lay outside their domains. Yet, the very products that they develop *de facto* create a food policy. And, who has the expertise in these areas if not those in the scientific community?
5. An additional problem arises in the social mechanisms necessary to maintain *in vitro* systems. Even today, with all food produced in the field, there are occasional documented cases of contamination of food products, either by accident, by a misguided or deranged individual, or even for political reasons. However, the production of food *in vitro* creates much greater security problems. Contamination of large quantities of food – for whatever reason – is easier when there is *in vitro* production, simply because so much food is produced in so small an area. Moreover, protection of such production facilities would be difficult and would necessitate undemocratic institutional forms. Given that some observers have encouraged *in vitro* production as a remedy for political turmoil, this problem presents both ironies and contradictions.

Conclusions

There is a certain irony in that the technological optimism described above is associated with social pessimism. However, as we have noted, even technological optimism must be examined carefully. This is – despite impressions to the contrary – an extremely difficult task. The only source of information about future directions in biotechnological research is the scientific community itself. That community is strongly biased in the direction of technological optimism. For example, some years ago scientists were infatuated with the prospect of modifying plant materials through irradiation; a little later we were told that the sea would provide the abundant food needed for future generations.

Neither of these predictions have come to pass. Current forecasts for *in vitro* production pay little attention to the intricate and expensive basic science that must still be conducted in order to modify even the most simple of plant characters (including unravelling the development of flavour in the edible parts), the complex developmental research necessary to arrive at production prototypes, and the economics of large-scale production. These are likely to be formidable barriers to large-scale production. However, even though the prognostications scientists provide may be biased, we ignore them at our peril. The possibility exists that a radical restructuring of what we now call agriculture could be triggered by these technologies. We must examine the likely consequences now and ask whether we wish to pursue this path. Yet, at the same time, we must remain skeptics.

On the other hand, the new biotechnologies offer us an opportunity to assess new technology before it actually exists. This is particularly desirable in that it permits – at least in principle – the avoidance of deleterious effects of proposed technical change through the reorganization of markets and creation of governmental and intergovernmental policies to direct that change. Nevertheless, much needs to be done. A more careful review and monitoring of the scientific literature is essential. In addition, more detailed forecasts of the impact of particular technical changes are needed.

Nevertheless, we can make some fairly strong statements about impacts if we accept the "optimistic" position that scientists provide to us. First, we can expect a major transformation of agriculture over the next quarter century. Specifically, we can expect the not-so- gradual elimination of spatial, temporal, and climatic barriers to food and fibre production. This alone will bring with it substantial social upheavals as the location of production changes. In addition, we can expect the elimination of major portions of the farming enterprise as many field crops are grown *in vitro*. This, in turn, will displace farmers and farmworkers on a scale never before possible. Moreover, most of these tens or perhaps hundreds of millions of persons will not be able to find work in other sectors of the world economy. Thus, the demise of agriculture would deprive many of the very means of subsistence. Another aspect of *in vitro* production is the enormous concentration of economic and therefore political power that it would create. A few firms and a few individuals would have control over the food supply. Concentrated production would also bring with it the possibility of deliberate or accidental contamination of the food supply; in a word, it would actually reduce our food security. And, finally, the human diet would be radically altered. The many "impurities" that are currently found in food would be removed, leaving us vulnerable to chronic if not acute dietary problems.

In papers of this sort it is typical to have some suggestions for resolving some or all of the problems raised in the paper. At first, it appeared that a New International Scientific Order might be called for. However, given the failure of the New International Economic Order, this seems most unlikely. More likely is that most of the nations of the world will ignore the problems raised herein as fanciful when compared to the very real problems of daily governance. Perhaps the most we can expect is a slow transformation that will still provide time for significant positive social change.

Finally, in closing, it appears appropriate to remember that the annual worldwide budget for biotechnological research does not equal what the nations of the world spend in a week on what is euphemistically called national security. Sad to say, it appears that the current definition of national security does not include people's right to food. If it did, the future would look much brighter indeed.

Notes

1. For a more detailed review of both conventional and new techniques of plant improvement, see Hansen et al, 1986.
2. Wittwer (1986) has noted that tissue culture of pyrethrin has made it possible once again to grow this insecticidal plant cheaper than to synthesize the insecticidal chemical in a laboratory. This should give us cause to withhold any simplistic generalizations about the role of tissue culture in replacing crops in the field.

References

Anderson, L.A.; J.D. Phillipson; and M.F. Roberts. "Biosynthesis of secondary Products on Cell Cultures of Higher Plants," pp. 1-36 in *Plant Cell Culture,* Advances in Biochemical Engineering/Biotechnology. No. 31. Berlin: Springer Verlag,1985.

Balandrin, M.F.; J.A. Klocke; E.S. Wurtele; and W.H. Bollinger. "Natural plant Chemicals: Sources of Industrial and Medicinal Materials," *Science* 228:, 1985, 1154-1160.

Baltimore, David. "Priorities in Biotechnology," pp. 30-37 in *Priorities in Biotechnology Research for International Development.* Washington: National Academy Press, 1982.

Boyer, Charles D. "Genetic Engineering: Tomorrow's Technology," American Vegetable Grower. 32 (April, 1984): 51.

Breuling, M.; A.W. Alfermann; and E. Reinhard. "Cultivation of Cell Cultures of Berberis wilsonae in 20-1 Airlift Bioreactors," *Plant Cell Reports*, 1985, 4:220-23.

Buttel, Fred H. and Randolph Barker. "Emerging Agricultural Biotechnologies, Public Policy, and Implications for Third World Agriculture," *American Journal of*

Agricultural Economics 67 (December, 1985):1170-75.

Bye, Pascal and Alain Mounier. Les Futurs Alimentaires et Energetiques des Biotechnologies," *Economies et Sociétés*. Hors Serie No. 27, 1984.

Chamala, S. "Transfer of Rubber Technology Among Smallholders in Malaysia and Indonesia: A Sociological Analysis," pp.20-34 in *Smallholder Rubber Production and Policies*. Canberra: Australian Centre for International Agricultural Research, 1985.

Chaleff, R.S. "Isolation of Agronomically Useful Mutants from Plant Cell Cultures," Science 219 (11 February, 1983): 676-82.

Chatelin, Yvon. "La Science et le Development: L'Histoire peut-elle recommencer?" *Revue Tiers Monde* 27 (105), 1985, :5-24.

Chatelin, Yvon and R. Arvanitis. Pratiques et Politiques Scientifiques, *Actes du Forum des 6 et 7 fevrier*. Paris: ORSTOM, 1984.

Clairmonte, Frederick F. and John Cavanaugh. "Destruction of the Sugar Industry," *Economic and Political Weekly*, 21 (4 January, 1986): 18-19.

Colin, H.A. and M. Watts. "Flavour Production in Culture," pp. 729-47 in D.A. Evans, W.R. Sharp, P.V. Ammirato, and Y. Yamada, eds., *Handbook of Plant Cell Culture*. New York: Macmillan, Vol. I, 1983.

Comis, Don and Marcia Wood. "Assembly Line Plants Take Root," *Agricultural Research* 34 (April), 1986,: 6-11.

Curtin, Mary Ellen. 1983. "Harvesting Profitable Products from Plant Tissue Culture," *Bio/Technology* 1(October, 1983):649-52, 54, 56-7.

Day, Peter. The Biotechnology Business: A Strategic Analysis. London: Frances Pinter, 1985.

Dembo, David; Clarence Dias; and Ward Morehouse. "The Biorevolution and the Third World," *Third World Affairs* I, 1985:311-25.

Dixon, Bernard. 1984. "Chemicals from Algae and Plant Tissue," *Bio/technology* 2 (August, 1984): 665.

Fisher, Arthur. "Science Newsfront," *Popular Science* 228(May,1986):10.

Food and Agriculture Organization. *FAO Trade Yearbook*, 1984. Rome: FAO, 1985.

Fowler, M.W. "Large-scale cultures of Cells in Suspension," pp. 167-74 in Indra K. Vasil, ed. *Cell Culture and Somatic Cell Genetics of Plants*. New York: Academic Press, 1984.

Gautheret, Roger J. "Plant Tissue Culture: A History," *The Botanical Magazine* (Tokyo), 96: 1983, 393-410.

Goldstein, Daniel. "Molecular biology and the Preservation of Plant Germplasm," Paper presented at the annual meetings of the American Association for the Advancement of Science, 1986.

Hansen, Michael; Lawrence Busch, Jeffrey Burkhardt; William B. Lacy; and Laura Lacy. "Plant breeding and the New Biotechnologies," *Bioscience* 36 (January), 1986: 29-39.

Heinstein, Peter F. "Future Approaches to the Formation of Secondary Natural Products in Plant Cell Suspension Cultures," *Journal of Natural Products* 48(1), 1985: 1-9.

James, A.T. "Plant Tissue Culture: Achievements and Prospects", *Proceedings of the Royal Society of London*, 222,: 1984, 135-45.

Kenney, Martin; Frederick H. Buttel; and Jack Kloppenburg, jr. "Impact of Industrial Applications: Socioeconomic impact of product dislocations," *Advance Technology Alert System Bulletin* 1, 1984, 48-51.

Lenin, V.I. *Theory of the Agrarian Question*. New York: International Publishers, 1938.

Martin-Leake, Hugh. "An Historical Memoir of the Indigo Industry in Bihar," *Economic Botany* 29 (October-December, 1975): 361-71.

Misawa, Masanru. "Production of Useful Plant Metabolites," pp. 59-88 in *Plant Cell Culture*. Advances in Biochemical Engineering/Biotechnology. No. 31. Berlin :

Springer Verlag, 1985.

Rogoff, Martin and Stephen L. Rawlins. "Food Security. A Technological Alternative." Unpublished paper, 1986.

Rosevear, A. and C.A. Lambe. "Immobilized Plant Cells," pp. 37-58 in *Plant Cell Culture*. Advances in Biochemical Engineering/Biotechnology. No. 31. Berlin : Springer, Verlag, 1985 .

Schaeffer, Gideon W. and F.T. Sharpe, Jr. "Mutations and Cell Selections: Genetic Variations for Improved Protein in Rice," pp. 237-54 in Lowell D. Owens, ed., *Genetic Engineering: Applications to Agriculture*. Totawa, NJ: Rowman and Allanheld, 1983.

Simon, Ruth. "Mail-order enzymes," *Forbes* 137 (11):1986, 46.

Staba, John. 1985. "Milestones in Plant Tissue Culture Systems for the Production of Secondary Products," *Journal of Natural Products* 48(2), 1985: 203-9.

Stanley, D.W. 1986. "Chemical and Structural Determinants of Texture of Fabricated Foods," *Food Technology* (March), 1986, 65-68, 76.

Tsai, D.H. and J.E. Kinsella. "Initiation and Growth of Callus and Cell Suspensions of Theobroma Cacao L." *Annals of Botany* 48, 1981, 459-557.

Van den Doel, Kees and Gerd Junne. "Product Substitution Through Biotechnology: Impact on the Third World," *Trends in Biotechnology* 4 (April, 1986).

Vergopoulos, Kostas. "The End of Agribusiness or the Emergence of Biotechnology," *International Social Science Journal* 37(3), 1985, 285-299.

Wittwer, Sylvan H. 1986. "New Technology Needed to Sustain Increased Food Production," In *Food for the Future*. Philadelphia: Philadelphia Society for the Promotion of Agriculture, forthcoming, 1986.

Yamada, Yasuyuki. "Selection of Cell Lines for High Yields of Secondary Metabolites," pp. 629-636 in Indra K. Vasil, ed. *Cell Culture and Somatic Cell Genetics of Plants*, 1984, New York: Academic Press.

Yamada, Yasuyuki and Y. Fujita. "Production of Useful Compounds in Culture," pp. 717-28 in D.A. Evans, W.R. Sharp, P.V. Ammirato, and Y. Yamada, eds., *Handbook of Plant Cell Culture*. New York: Macmillan, Vol. I, 1983.

Zenk, Meinhart. "The Impact of Plant Cell Culture on Industry," pp. 1-13 in Trevor A. Thorpe, ed., *Frontiers of Plant Tissue Culture*. Calgary: International society for Plant Tissue Culture, 1978.

2

BIOTECHNOLOGY AND DEVELOPMENT OF AGRICULTURE IN THIRD WORLD COUNTRIES

The Next Phase of Technology Transfer*

SHRIPAD DEO & NAVID MOHSENI

Introduction

"Revolution" is a word which has been bandied about routinely in recent years. We have heard it often enough in agriculture. First there was mechanical revolution, then the chemical, and in not too distant past, the green revolution. There is a purpose to the choice of this word. It attracts everyone's attention, and since there are no bloody skirmishes, seizing of palaces, and dethroning of monarchs and curtains ("iron", "bamboo" or "sugarcane") falling over the existing political orders, everyone feels relieved. Alas, it is a silent, non-violent and peaceful revolution. "Biorevolution", the advent of biotechnology and genetic engineering in agriculture, which could mark the beginning of a new phase in the way man has cultivated food and non-food crops and raised animals in the past, is of this genre.

Notwithstanding all the euphoria over "biorevolution", biotechnology is not new. Man has been using biologically based processes since the beginning of civilization, if not earlier. Wine, beer, cheese, bread, and yoghurt are all examples of foodstuffs in which a processing step mediated by a microorganism is essential[1]. Throughout history, even the poorest farmer has been an enterprising plant breeder, perpetually improving his seed, picking out the healthiest and fattest plants as

* We wish to thank Lahry Busch, Dwight Billings and Lou Swanson for their comments and suggestions on an earlier draft of this paper. We appreciate all of the help of Alessandro Bonanno in getting this paper ready for publication. The responsibilities for the final product, however, is only ours.

parents for the next season's crops. This continuous process is still at the heart of agriculture throughout the world, both developed and underdeveloped. The emergence of biotechnology and genetic engineering is now giving a new meaning to this process[2].

Then what is it that makes this new advent of technology an object of everyone's attention? First, it is still nascent at the moment and full of promises which can make science fiction books read like history. Second, we hear the familiar arguments of the green revolution years, now being made in support of biotechnology like, "this technology will improve human diet and increase production," or "an increase in agricultural production will improve general conditions both in developed and developing countries. "Implicit assumptions of these arguments are (a) social, economic, and political problems can be solved through newer technologies; (b) technology is neutral among users; and (c) increased production, not distribution of wealth, is the problem. Third, this new "revolution" is managed and executed by corporate capital and government organizations in developed industrialized nations with the restructuring of universities, laboratories, industries and companies, and even laws. Fourth, the Gadarene rush for acquisition, development and control of biotechnology in developed nations has also caught on with underdeveloped nations without much prodding.

It is in this context that the adventure of biotechnology is seen. Biotechnology and genetic engineering have application in many fields like industry, medicine, energy, waste management, and agriculture. This paper will concentrate on application of these new "high technologies" to agriculture. The first section deals with historical and theoretical aspects of science and technology, that is, what is the nature of technology, its appropriateness, and the social aspects of technology, for example, how technology is produced and used, who are the beneficiaries, and so on. The second section looks at agricultural research, economic development, and the transfer of technology in historical context. The third section ideals with the advent of biotechnology, specifically in agriculture, and its potential effects on agriculture in Third World countries. The final section attempts to assess the impact of new technologies on underdeveloped countries in terms of their existing dependency and changes that are likely to occur as the technology develops further. It also examines the role of International Agricultural Research Centers (IARCs) and the proposed International Center for Genetic Engineering and Biotechnology (ICGEB) in development, control and access to this new technology.

Science and Technology

Historical and Theoretical Perspective

"The Party views scientific and technical progress as the main lever for the solution of all economic and social issues" (Mikhail Gorbachev, June 1985): "No nation depends as much as we do on the science base" (Ronald Reagan, February 1985): "Science and technology have emerged as a universal language for humankind" (Yasuhiro Nakasone, June 1985)[3]: and "Scientific endeavour... instills confidence in society and leads it to a higher sense of achievement and fulfilment...scientists are aware that new knowledge is often the best way of dealing with old problems" (Indira Gandhi, July 1982). These pronouncements by political leaders of our time provide a seductively simple explanation for complex social, economic and political problems. They are also sufficient to convince anyone of the dominant role of science and science-based technology in our society today. The developed industrial nations need it to maintain their domination in the world economy and the underdeveloped nations look at it as a means of survival, transformation, and modernization. It is, therefore, necessary to examine the exact nature of science and technology, and its precise interconnections with social reality.

The knowledge of nature which human beings have is a raw material for science. "Nature not only exists in itself, but also, and more crucially it becomes 'for-us' in so far as nature is drawn into the web of human and social purposes."[5] This knowledge of nature is accepted in its broadest sense for the process of producing a particular form of knowledge. The outcome of this transformative activity of scientific practice is the construction and reconstruction of systematic explanations of observed natural and social phenomena. It is arguable that not explanation alone, but prediction, control, and description are also the result of scientific activity. But, prediction, control, and description can follow from explanation, not vice versa.

The process of creation of scientific research is not a function of individual creative abilities alone. To put it more accurately, this process needs to be placed historically within a specific, dynamic, and differentiated social reality.[6] The development of social research reflects the character and level of society's development and particular social relations within it. It is possible for scientific and technological activities to exist alongside and interpenetrate each other. Technological activity is distinct from scientific practice and is characterized by its drive for the mastery and control of nature to serve the purposes of human beings. But this drive is distinct from a drive to understand nature.[7] It is this distinction which is getting blurred in the context of biotechnology.

The notion of "domination of nature" is very old with its roots in the Judeo-Christian tradition. In every human culture there are tendencies for direction of social groups toward the world of nonhuman nature. These patterns determine the ways in which the nonhuman nature may be allocated for the satisfaction of human needs.

> Western capitalism was the first social form with a growing tendency to regard all nonhuman nature in exclusively utilitarian terms as being nothing but a warehouse of resources for human use, and to systematically denigrate any other symbolic significance in the human relationships to it. In other words, there arose a vision of nonhuman nature as mere 'material' that could be subjected to complete and unchallengeable control through the progress of human scientific and technical ingenuity.[8]

In this context it needs to be emphasized that science per se does not provide the human race with either the means of survival or meaning for enduring hardships. The new knowledge and know-how have served the privileged of this world to the detriment of the rest.[9] It can be argued that the great cultures of Asia would have been able to absorb, nourish, and reproduce the first Western innovations (in the sense of civilization) had they been introduced under suitable conditions. The same could also be said of Africa and Latin America. "The privilege on which the West can presume could have been shared by the whole world if decisive historical factors had not led to a different result. Science...was regionalized by the history of its dissemination.[10]

Science has become a direct productive force permeating and transforming all areas of human existence. It is deeply rooted in institutions, intellectual life, public policy, and technology. Science also aids the capitalist mode of production in its continual need for innovation in all walks of life, the creation of new commodities, new technologies, new ideas, and new social structures. The subordination of science or the scientific world view to the capitalist mode of production is an "ideological representation–or even a legitimation–of the bourgeois world, but...as the intellectual tool which capitalists use to organize, plan and control the labour process, it is, in fact, a constituted element of the world. And...the need for tool has not been determined by the requirements to increase the 'efficiency' of production in any neutral sense, but in addition by the need of capital to impose such organization, planning, and control over labour, in order to secure the generation and subsequent appropriation of surplus value."[11]

This scientific world-view which was a liberating force in the West became a means of domination, creating dependent intelligentsia and culture in the Third World nations. To these nations trapped in the network of trade and commerce, science was known by its products long before it could be known for what it was. By introducing consumption of knowledge long before its production, the local intellectuals and social leaders were in a situation where they were forced to copy the ready-made goods before they were taught how to produce those which were appropriate to their needs.[12]

Science and technology have been transferred to the Third World under conditions dictated by the desire to enlarge those markets which are likely to make investments profitable both locally and in places from which they originate. This technology not only affects the nature and structure of society, but it is also strongly influenced by socio–economic systems. Technology is not merely a mode of production, hence, universal and neutral, but it also carries with it, to use the genetic metaphor, a "code" of structures–economic, social, cultural and cognitive. The economic code that exists in western technology (again in the sense of civilization) demands that industries be capital-intensive, research-intensive, organization-intensive, information-intensive and importantly, labour-extensive. On the social level, this technology has a tendency to create not only a 'centre' and a 'periphery' but also to perpetuate a structure of inequality and dependence.[13]

The transfer of this technology then takes on the shape of cultural-structural invasion which is obfuscated by the fact that it is not accompanied by the colonial presence as in the past and partly because "the fragmentation inherent in Western technology fragments the perception of the total picture.[14] Third World countries strive desperately to catch up with developed countries by means of technology and place their hopes in technologies as a way to improve living conditions. In their efforts to free themselves from neo-colonialism, the imported technologies make them more dependent on developed industrial nations. Agriculture is the case in point.

Agricultural Research, Development and Technology Transfer

A Brief Historical Review

Historically, farming in Third World countries was not geared to generate large-scale surpluses, both real and financial, which provided the "nutrient base" for the development of nonagricultural sectors. To understand the capitalist penetration of agriculture, it is necessary to differentiate between farming and agriculture.[15] Farming is a process of turning seeds, manure or fertilizer and water into rice, potatoes, corn and cotton using land, human and animal labour and machines on the farm. Agriculture includes farming but it also includes all those processes that go into producing, processing, transporting and marketing purchased inputs like seed, machinery and agro-chemicals by the farmers, and final product which consumer buys in the store.

To understand the role of agricultural research in the expansion of the capitalist mode of production and increasing concentration of capital that is almost a global phenomenon, it is necessary to go back to the period when the European nations went to great lengths to maintain their control over plant resources they encountered in their journeys overseas. Their initial systematic efforts materialized in the form of botanical gardens which were spawned in Europe and later in the colonies. These botanical gardens were depots for acquisition, improvement and transfer of plant material with economic possibilities from one part of the world to another where it could be raised on plantations. This phase of specialized and systematic exploration of plant resources, or agricultural research in its rudimentary form, led to expansion of colonial empires based on, among other things, "development of several highly profitable and strategically important plant-based industries in tropical colonies."[16] It is here that the initial link between agricultural research and the burgeoning world capitalist economy is seen, or as Brockway puts it, "There is no way to draw the line between science, commerce and imperialism" in the work of botanical gardens.[17] This link becomes clearer in the context of biotechnology discussed in the next section.

The agricultural development that took place in the colonies was oriented towards cash crops like sugar, jute, rubber, coffee, cocoa, tea, and bananas. The emphasis was on rapidly increasing the production on well-endowed lands. A vast majority of farmers were subsistence farmers who were more concerned with maintenance of the small plot of land to produce enough for their own consumption and for payment of taxes. Although these farmers, within their own methods of cultivation, were rational and efficient, when compared to the Western agricultural producers and economic cost–benefit criteria, they were cast in unfavourable light.[18]

Agricultural research in developed countries progressed beyond taxonomy of commercially beneficial plants and foodcrops. The colonial botanical gardens functioned as agricultural experiment stations performing the developmental work by focusing upon testing varieties in new climates and devising new methods of cultivation.[19] It was in the 1850s that von Liebig's work in agricultural chemistry linking minerals in soil to the growth of plants led to research in Europe on chemical fertilizers as nutrients. This subsequently enabled farmers to increase yields per hectare. Agriculture was on the way to becoming a "science." The first agricultural experiment stations were established in the 1850s in Germany and soon followed by France, Britain, Russia, Japan and the U.S. Between 1900 and 1930, there were more than 1,400 stations worldwide. The expansion of agricultural research did not promise an immediate source of profit; hence, most of them were supported by the state. These stations outside Europe directed their research toward export-oriented crops while in Europe it was focused on foodcrops.[20] This situation remained unaltered until the end of World War II.

It is obvious that agricultural research had a different orientation in developed countries and in the colonies. In the "core" countries it served the needs of the state and "shared the mercantilist and nationalist spirit of the time,"[21] while in the colonies it was forced by the colonizers. The concentration of exportable crops and compulsory export of available foodgrain, leven during famine years, put many Third World countries at a disadvantageous position in terms of improving foodcrops.[22] Another outcome of this research system was that it produced and trained agricultural scientists who tended to think in the same way as the researchers from developed countries did.[23] This trend is observed even after independence. The output of this agricultural research system was not appropriate to fulfil the needs of the Third World countries, but it brought about structural changes which have fundamentally and irreversably altered the traditional societies.[24]

The development of Europe during the last 150 years beginning with the Industrial Revolution in Britain, based on the growth of science and technology and spread of industrialization, has provided an attractive model for underdeveloped countries. This science and technology not only increased the available surplus of societies that gave birth to it, but also increased both quantity and quality of goods and services available to their people. This perceived potential of technology to provide an apparent unending supply of commodities led most of the Third World countries to follow the same road. The experiences of developed industrial nations of recovery following the depression of the 1930s in the U.S. and reconstruction of war-torn economies of Europe suggested a model of growth which emphasized urbanization, industrialization, and division of labour in agriculture and industry.

The technology that was offered for industrialization was capital-intensive and labour displacing. The large amount of capital required was not always available because the traditional agricultural sector was not oriented toward production of surplus on a scale which would facilitate large-scale rapid industrialization. The attention given to the development of commercially important crops at the expense of food crops for centuries during the colonial period resulted in an agriculture which could neither provide food nor could it generate a large enough surplus for industrialization. It was assumed that lack of food is the cause of hunger, persistent poverty, and resultant rural unrest; hence, increase in food production was the way out. What was needed was the transfer of technology which would increase yields dramatically, not land redistribution and other agrarian reforms.

An increase in agriculture production, in lieu of land redistribution and other agrarian reforms, was suggested as way of overcoming the problem.[25] "Yield increases of 15 to 20 per cent will convince no one. Demonstrations showing increases from 100 to 600 per cent...have caught the peasant's imagination and have triggered off agricultural revolution," wrote Norman Borlaug, pioneer of the "green revolution" and winner of Nobel Peace Prize.[26] The agricultural technology that promised spectacular increases in yields advocated the use of seeds of cereals like rice, wheat and maize which were highly responsive to chemical fertilizers; chemical fertilizers, water management, pesticides, and mechanization.

This transfer of technology for "modernization" of agriculture in underdeveloped countries, better known as "green revolution", represents the most recent large-scale, systematic, and coordinated effort in transferring agricultural technology from developed nations, mainly the U.S. and the IARCs, to the Third World nations. The new varieties of seeds developed by IARCs in Mexico and the Philippines had many built-in characteristics, namely, high response to agrochemical inputs and irrigation; their dwarf and semi-dwarf size ensured that a large proportion of nutrients went into the production of grain, their inbred insensitivity to light enabled them to mature within a given time period facilitating multiple crops, and so on. The promise of three-to four-fold increase under laboratory conditions which these varieties showed could not be transferred to the fields where conditions and available inputs were often different.

The new technology "package" could deliver high yields only if used according to recommended practices. In the Third World countries where a large proportion of farmers are accustomed to a "self-sustaining local production/consumption system", they now have to depend upon inputs including seeds, and information developed and disseminated from outside. Each aspect of this technology requires larger cash investments than those required under traditional agriculture. The creditworthiness of farmers becomes crucial in obtaining loans from financial institutions. Small and marginal farmers thus have and inherent disadvantage in access to credit on realistic terms. Besides many resources are allocated by the state bureaucracy at a price below a market clearing price or no charge at all. They also define who is a small farmer or who is eligible to receive loans or subsidies through state-controlled organizations. Under these circumstances the crucial question is: Who has access to the institutions which produce and deliver these services and on what terms?[27] In short, the farmer has to reorganize his economy to produce a minimum surplus to pay for the inputs and interest due on the initial loan.

The inbred insensitivity to sunlight which enabled farmers to grow multiple crops in a year also meant that labour requirements for land preparation, sowing, and harvesting would increase. But increased dependence on labour to fully exploit the potential of new seeds shifted the leverage in favour of farm labour. This inconvenience was overcome by resorting to mechanization of various operations. A potential for increased employment in rural areas was rejected in favour of more control through mechanization. The preference for mechanization also meant that small landholders were operating at uneconomic level. The bias of capital-intensive technology in a labour-surplus economy manifested itself in many ways.[28]

These new varieties not only increased yields in regions where conditions were suitable but also raised a spectre of potential epidemic zones. The new varieties were prone to disease and pest infestation since they were developed in laboratories and were not a product of natural selection. Their narrow genetic base has created an alarming uniformity making them vulnerable to diseases and pests.[29] This problem has already reached the critical stage in developed countries and has caused alarm in Third World countries. The rapid increase in acreage under new varieties in these countries has also led to reduction in area under traditional varieties, resulting in the erosion of genetic diversity.[30] The use of chemicals to control weeds, pests, and insects has resulted in environmental pollution on a large scale.

The erosion of genetic diversity has created concern about availability of genetic resources for the future breeding of both crops and livestock. It also leads to the important question of access, preservation, and availability of germplasm for future plant breeding. The wild strains which have not yet been genetically altered by domestication and hybridization are required for plant improvement. This is an essential ingredient for new technologies like biotechnology and genetic engineering. Without this raw material new varieties or products through genetic manipulation are not possible. The centres of genetic diversity, the Vavilov Centers, which are essentially located in the tropics, have been the sources of almost all of world's basic food groups. This explains the rush for acquisition of and access to germplasm by various developed nations and MNCs based in these countries.[31]

The emergence of biotechnology and genetic engineering, especially in agriculture, needs to be situated in this context. It is also necessary to examine this development in the context of the new international division of labour and changing nature of dependency.

Development of Biotechnology in Agriculture in the Changed International Context

To understand the emergence of biotechnology, it is necessary albeit briefly (for reasons of space, rather than its importance) to look at the international economic situation. The last decade and a half has been characterized by a prolonged crisis in the capitalist world economy. Various indicators like the fall in the rates of overall economic growth; low rate of capacity utilization; drop or stagnation in investment in industrial plants in developed countries; change in the structure of the international division of labour; rapid spread of production facilities in the Third World countries oriented towards the world market; growing synchronization of business cycles which makes it difficult to take effective anti-cyclical measures, based on the internationally unsynchronized nature of national business cycles (consider the outcomes of the so-called "world economic summits" in the recent past) to prevent economic crises; radical redistribution of world incomes after oil-crisis is manifested in the changed structure of world trade; increased balance of payments problems and debts for many developing countries; rising or stagnating unemployment at a relatively high level; growing disparity between skill-structure of those seeking work and vacancies; increase in intensity of conflict between employers and workers over maintenance of real wages, jobs, and conditions in developed countries, and so on; point to the fact that the capitalist world economy is at a turning point. In short, the post-war boom experienced by the world economy is coming to an end and with it the search for new methods for the transforming structural features which kept capitalist world economy stable is on.[32] The structure of the international division of labour has changed, enabling relocation of production of manufactured goods from industrial countries to developing countries which are competitive on the world market. The result is widespread economic dislocation, unemployment, and socio-economic disintegration of old industrial regions. The need to encourage "high technology" industries felt by almost all industrial nations has "precipitated unprecedented technological competition" among these countries. "Biotechnology is, of course, one of the most prominent technology... and has been the object of significant public R & D subsidies in most industrial countries.[33] Micro-electronics, satellite communications and remote sensing, and fibre optics are some of the other high technologies. As the competition among the nations of North America, Western Europe and Japan increased for the development of these new technologies, the threat to U.S. technological domination loomed larger. The relocation of manufacturing industries in developing countries has benefited some of them, contributing to both intense growth and substantial debts. The development of industries like textiles, footwear, steel, rubber, electronics, ships, automobiles and auto parts, glass and chemicals in Newly Industrialized Countries (NICs) like Brazil, Mexico, South Korea, Taiwan and some centrally planned economies is on the rise.

In agriculture, the green revolution technology had proved to be less than revolutionary. It was spatially uneven in its applicability and diffusion. The expansion in acreage under improved varieties of rice and wheat has taken place in the more favourable agro-ecological zones and where the infrastructure was suitable. As the technology was spread to marginal lands, the yields began to drop or stagnate. The post-oil crisis period also showed that the high-cost, capital-intensive technology was burdensome, especially to small farmers in developing countries. In many of these countries it led to polarization of the rural areas into rich capitalist farmers and the rapidly rising ranks of rural proletariat.[34] Declining yields, high cost of inputs, and dwindling state support led many of the early beneficiaries of green revolution to world market oriented production in these countries (e.g. fruit, vegetables, flowers, meat, etc.). This shift is supported and financed by international financial agencies.[35] Another important factor is the agribusiness. The agribusiness corporations attempt to wed developing country needs to their owns "interest and capabilities" and they believe that they alone are capable, qualified, and have sufficient expertise to develop Third World agriculture and feed the world.[36] The needs of Third World countries are identified with the profit-earning capacity and power of these corporations. In order to ensure their dominance over agriculture in Third World countries, agribusinesses strive for control to decide what, where, when, how and how much to produce; whether, when, how, and to what markets to distribute the production, and how to assure and maximize the transfer of profits from these countries. This control is also to be seen in the developed countries.[37] The transfer of capital is followed by technology transfer because the inherent logic of the system is that this will be followed by the transfers. As discussed above, it is intrinsic to the nature of technology that is transfered.

It is not happenstance that agribusinesses, chemical and pharmaceutical MNCs, got interested in biotechnology and genetic engineering in a big way. As mentioned above, biotechnology is a set of techniques that involves the use of biological processes and living organisms in the industry, agriculture, and other areas of economic activity. It has been dichotomized into "new" and "old" techniques. The former category includes processes, like recombinant DNA (rDNA), cell fusion, etc. while the latter category involves traditional methods like fermentation, use of enzymes in biological reactions, and so on. It is important to note that the cluster of new techniques, called biotechnology, is based on biochemistry, genetics, cellular biology and molecular biology and especially applied molecular biology. It was through the financial support of the Rockefeller Foundation that interest in molecular biology got a boost, but it remained within the confines of "basic" research until 1970s when Cohen and Boyer made a spectacular discovery that it is not only possible to insert foreign DNA into a living bacterial cell, but also induce "expression" of the foreign DNA's instructions. In short, it is possible to "program" nature and make it work according to human instructions.[38]

This discovery was notable in many ways. It was a significant step in man's quest for the "domination of nature". It opened up a new prospect for "practical" applications in many fields and there ensued a fierce competition for funds and access to new information. Cohen and Boyer obtained a patent for their technique and the "Diamond vs Chakravarty" decision by the U.S. Supreme Court in 1980 indicated the directions in which the new industry would develop. In the process scientific research was redirected, re-aligned and rearranged in ways which facilitated the control of 'new' biology by capital.[39] Among the significant aspects of this technological development are the increasing privatization of research, participation of universities and university-based researchers in the development of products and processes which will enable the burgeoning biotechnology industry to expand.

In most industrialized nations research and development in biotechnology is carried out by universities, small biotechnology venture capital firms, and MNCs. But in most instances those engaged in research get support from governments. Traditionally, universities and other publicly funded organizations concentrated on basic research. In the present situation of severe competition among researchers for grants and among universities for funds from MNCs with prospects for obtaining funds from research results to support activities in the universities, science and scientific research has reached a point where "conception *is* capitalization."[40] The increasing linkage between universities and industry has raised problems of conflict of interest and importantly, of transforming 'public' research into a "private" property. The venture-capital firms are small in size and have low levels of capitalization, this has tended to put limits on their ability to upscale the processes they develop and have restricted their activities, thus far, to research, consultancy, and services. It is the MNCs who have not only developed their own research and development facilities but have also established strong linkages with universities and venture capital firms.[41] It is not restricted to U.S. based MNCs, because many Japanese companies have also struck numerous agreements with venture capital firms. "If it wasn't for Japan, a lot of entrepreneurs in the

U.S.A. wouldn't have made it".[42] That has many Americans worried because the extent of technology transfer to Japan is not known. Besides, when the biotechnology industry itself is looking for marketable products, Japan has shown an ability to organize industry for exploitation of fundamental research. The Japanese biotechnology industry has introduced six times as many drugs per dollar spent on research and development as did comparable U.S. companies.[43] By far, the U.S. and Japan have taken a significant lead over other industrialized nations in terms of research and development. There is a significant level of difference in allocation of funds and progress made in research and development in biotechnology in various industrialized countries, with many resigned to the reality that they will be able to concentrate only on a few areas.

Agriculture as a way of life and as an occupation for gainful employment for millions of farmers is threatened by the way agriculture is transformed into industrial processes. The prospect of agricultural production in the field being supplanted by plant cell and tissue culture is a distinct possibility.[44] The impact of agriculture in Third World countries will be profound. It is not only that the commodities traded by these countries, like sugar, coffee, tea, cocoa, rubber, cotton, oils, spices, and other aromatics will be affected by substitutes available through biotechnology. The control of improved and tailored varieties of seeds will now be available through MNCs. How quickly MNCs consolidate their hold on the seed market globally is important because it will push many Third World countries toward deeper dependency. For example, it is argued by many proponents of biotechnology that it is possible to engineer crops resistant to many herbicides, which will reduce production costs as well as the need for tillage. The global sales of herbicides in 1984 were almost $4 billion. It is highly unlikely that the new herbicide-resistant varieties will reduce the usage of herbicides and kill the market. On the contrary, the agro-chemical MNCs are developing varieties that can withstand herbicides. This will lead farmers, who can afford it, to "flood their fields with more herbicides", leading to more pollution. The MNCs are designing plants in a way that will protect their market for specific herbicides that are already established in the market and whose patents might run out in the near future. "Genetic engineering allows us [Monsanto] to enter a partnership with mother nature. We don't have to force-fit herbicides to plants any more."[45]

In short, agriculture in Third World countries will be disrupted and transformed. It will affect not only the trade in various commodities but it will also lead to disruptions of population and migrations to urban centres. The scenario depicted by Rogoff and Rawlins, if it is made cost-effective, would lead to fields planted by biomass perennial crops. These are harvested as needed and are turned into sugar using enzymes. This sugar solution is then piped to plants located in urban or metropolitan areas where food is produced through scaled up tissue cultures using sugar solution as a medium and nutrient source for plant material. This will make processing a year-round activity with production restricted to what is needed on a particular day. This will eliminate canning and freezing processes as well as the spoilage. The production can be switched from one commodity to the other as the demand changes.[46] If this is not sufficient, development of agricultural robots capable of milking cows, pruning trees, harvesting, and picking; laser-guided tractors and other farm machinery are on-line to improve efficiency and reduce cost for the industry. It will relieve "people of the more tedious jobs they are required to do at present."[47] It is the production cost which is holding these 'exciting' developments in technology back How will agriculture in the Third World fare under this onslaught?

Impact of Biotechnology on Third World Agriculture : New Dependency?

In the post-war period underdeveloped countries have been enthusiastic about the notion that technology is a dynamic factor in economic growth, and hence, there is an urgent need to transfer technology from the developed countries. There is a fascination about the convenient assumption that developing countries being "latecomers to development" can use the available technology without following the costly course of "reinventing the wheel." Many studies have shown that the contradictions between self-reliant growth and technology transfer increase.[48] The transfer of technology to Third World countries is a one-way relationship which turns out to be detrimental to their interests. The control exercised by technology suppliers may deny the opportunity to Third World countries to adopt, generate and operate technology suitable to their development needs.

Technology is viewed as a commodity which can be bought and sold in the international markets with its imperfections. The usage of the term "technology transfer" tends to divorce the commercial and economic aspects of technology and hence, obfuscates its real nature as an instrument of domination.[49] As long as technology is embodied in physical goods, the market for it is not independent but is an integral part of the market for the goods. When these goods are sold as a package, the non-competitive conditions for each are created. The patenting system, trademarks, and other institutions of private property rights which cover the knowledge part of technology enable the seller to gain and consolidate monopoly power. This reality not only arrests growth of recipient countries prematurely but also accentuates the high income disparities prevalent in the socio-economic structure of these countries. The case of green revolution illustrates this point.

In the context of biotechnology, it is possible to show that its development is a continuation of the strategies used during the green revolution years.[50] The same belief in the potential of technology to bring about autonomous changes in the political, social, and economic structure is evident. The proponents of biotechnology would prefer to lock themselves in the "iron-cage of technology" so that they can get away with superficial examination of its far-reaching implications.

The domination of MNCs in biotechnology raises a number of points which Third World countries need to consider. The private nature of research and development enables them to have complete control over their processes and products. Their control over vast financial and physical resources and global markets and their ability to use leverage to manipulate State apparatuses makes it difficult to visualize the possibility that their priorities (maximize profits globally) will be in tune with those of the underdeveloped countries. The case of malaria vaccine provides a good illustration. Privatization also creates problems of access and leads to restrictions on free flow of research information, perpetuation of technological dependencies, inappropriately "packaged" technologies, and inequities in terms and conditions of availability of information. The vertical integration within MNCs and the patent laws enable them to enlarge the veil of secrecy around the knowledge-based technologies. This secrecy is not restricted to the research in their own organizations but also to the research funded in various universities. This problem is faced not by Third World countries alone but also by IARCs who played a leading role in the dissemination of green revolution. Considering the level of funding available to them, the comparatively limited facilities and staff available at IARCs and the competition between nations to capture the lead in the race, the IARCs "will probably limit themselves to the role of users rather than originators."[51] The IARCs, according to Donald Plucknett, senior scientific advisor at CGIAR, will enter into collaborative agreements with organizations specializing in biotechnology to use the new technology. But they will not commit their resources to "basic and very specialized research."[52] In this context the assertion that the private character of applied agricultural biotechnology "will indirectly benefit the IARCs by spreading the rate of technology transfer"[53] appears contentious. If the technology carries with it a code of structures specific to the social formation(s) of its origin, then biotechnology can only exacerbate the effects of green revolution technology.

An alternative to technology transfer for Third World nations is to be self-reliant in terms of research and development. To be self-reliant, it is argued that countries would need not only large enough scientific and technical manpower and sizable support from the governments. The classic model of a linear relationship between science and technology which implied that the existence of a scientific community is sufficient to ensure rapid economic growth from the introduction of new technology does not stand. For example, a critical assessment of the situation of the scientific and technical community in India shows that while a large-sized scientific community is at best a requirement for development, it certainly cannot be a sufficient requirement. It is not the size of the scientific community that really matters, but the quality.[54] The question of research priorities reflected in science and technology policies of these countries and dominant class interests need to be examined. Of its total budget for scientific research and development, India has allocated only six percent for agricultural research and 12.67 percent for industrial research. Atomic energy, space and defence claim almost half of it.[55] In addition the 1982 technological policy statement and the quest to take India into the 21st century in a hurry has opened a floodgate for foreign collaborations, turnkey projects, and scrapping of not only indigenous know-how but also basic research programmes.[56] If the class interests prevent indigenous research and development in a country like India, smaller nations with little or no infrastructure will have no opportunity to benefit from new technologies except through imports.

It has been argued[57] that the Newly Industrializing Countries (NICs) like India, Brazil, Mexico, and Argentina have the capacity to develop indigenous capability in biotechnology. The advantages of cooperation and unity among Third World nations will act as a check to domination by industrial nations. These arguments tend to overlook the fact that these NICs form an intermediate tier, Wallerstein's "semi-peripheral nations", between the exploiters and the exploited. These countries seek trade relations and technology from the developed countries, supplying them semi-processed goods, while obtaining raw materials from underdeveloped countries and exporting finished prodcuts to achieve intermediate wage levels and margins of profit. In the new international division of labour, these countries provide a relief or a safety-valve from the pressures on the developed countries as grateful recipients of profits from the wage-productivity squeeze. The relations between NICs and underdeveloped countries embody disturbing parallels to the relations between industrialized and developing states, an addition to the web of unequal international relations.[58] The short history of UNIDO's proposed International Center for Genetic Engineering and Biotechnology illustrates this point.[59] It is necessary to bear in mind that in analyzing the quest of Third World nations for biotechnology, it needs to be seen not in the mechanistic North-South relations but in the context of international class relations and the prospect of exploiting their existing and potential markets abroad.

Conclusion

If we accept that history is not a series of discontinuous events and that "the future is systematically connected to the past,"[60] then the advent of biotechnology in agriculture in particular can be seen as a continuation of old patterns. In spite of all the technological optimism, the direction of research and the benefits from the so-called biorevolution are not for the needy and poor farmers in Third World countries. They are not the appropriate market for the technology that is being amplified as new, significant, path-breaking, less capital-intensive, and more information-based. The existing class structures and class nature of states in NICs (other underdeveloped nations can only read about this technology as a horrifying science-fiction come true) which perpetuate and increase pauperization, can only ask the poor to accept their situation as something ordained by fate.

To argue that biotechnology has the potential to solve problems faced by agriculture in Third World nations is like setting up your own strawman and then showing how efficiently he can be destroyed. In spite of the vast literature critical of the green revolution, the advocates of the biorevolution have not been able to come up with any new technology to solve the problems which are essentially economic, political, and social which are deeply rooted in the existing structures. The increased concentration of economic and political power makes industrial nations feel it unnecessary to disguise their real intentions. Consider the following remarks by Nyle C. Brady, Senior Assistant Administrator for Science and Technology, Agency for International Development.

> Experience during the past two decades shows that American farm exports have generally benefited from increased agricultural productivity in the Third World. Countries such as Taiwan, Korea, Brazil, and Nigeria, which once were recipients of U.S. technical assistance, are now among the major purchasers of U.S. food exports. Developing countries have become the fastest growing markets for U.S. food exports... Collaborative research with Third World has benefited U.S. agriculture...through the infusion of yield-producing genetic materials into seeds of our cultivated crops...Continued scientific and technical assistance to the developing countries is essential and in the long run will provide expanded trade opportunities for U.S. agriculture and industry.[61]

This quotation leaves nothing to chance as to how biorevolution will shape the future for Third World countries. It is for these nations to decide whether their national plans for development of biotechnology are meant to be real catalysts of progress or cosmetic accoutrements for covering their poverty of political will to tackle what are essentially social, economic, and political problems, with biotechnological adornments.[62] The real question is how the millions of small farmers in Third World countries will react to the choice of their national leaders.

Notes

1. Jackson (1984), p.17 and Yoxen (1983), p. 14.
2. Mooney (1980) and Yoxen (1983).
3. Bloch (1986), p. 595.
4. Gandhi (1982), p. 1008.
5. Schmidt (1971), pp. 58-59.
6. Hessen (1930).
7. Leiss (1972), Part I.
8. Leiss (1977), pp. 131-2.
9. Moraze (1979), p. 255.
10. Moraze (1979), p.245.
11. Dickson (1979), p. 32 and also Rose and Rose (1976), pp. 14-31.

12. Moraze (1979), pp. 246-7 and Goonatilake (1984).
13. Reddy (1976), p. 40; Goonatilake (1984), p. 120-22 and Galtung (1979), p. 283.
14. Galtung (1979), p. 277.
15. Levins and Lewontin (1985), p. 210.
16. Brockway (1979), p. 6 and also Busch and Sachs (1981).
17. Brockway (1979), p. 84.
18. Dogra (1983), p. 222 and also Sen (1984), p. 166.
19. Brockway (1979), p. 165 and Busch and Sachs (1981), pp. 132-33.
20. Busch and Sachs (1981), pp. 133-36 and Pray (1984).
21. Brockway (1979), p. 75.
22. Spitz (1978), pp. 879-81.
23. Wellhausen (1975).
24. Goonatilake (1982 and 1984).
25. Joshi (1972); Ping (1979); and Pearse (1980).
26. Borlaug (1973).
27. Schaffer (1975).
28. Pearse (1980), Dasgupta (1977) and Byres (1978 and 1980).
29. Safeeulla (1977), pp. 72-85, Dogra (1983), pp. 222-224.
30. Mooney (1980); Reichert (1982) and Dogra (1983).
31. Mooney (1980 and 1983) and Geevan (1983).
32. Frobel (1980); Frobel, Heinnichs and Kreye (1980); Frank (1984); Buttel (1984) and Gereffi (1985).
33. Buttel (1984), p. 4; Dembo, Dias and Morehouse (1985); Morehouse (1985) and OTA (1984).
34. Cleaver (1972); Pearse (1980); Dahlberg (1979); Valianatos (19); Byres (1981) and Oasa (1984).
35. Feder (1979), p. 22.
36. Feder (1976), p. 40; Leopold (1985); Morgan (1979) and Valianatos (1976).
37. Vandermeer (1982) and Friedland, Barton and Thomas (1981).
38. Bull, Holt and Lilly (1982); Buttel and Kenney (1985); Zalinskas (1983); Lakoff (1984); Yoxen (1981 and 1983); Dembo, Dias and Morehouse (1985); Dickson (1984); Norman (1981); Doyle (1985); Hackling (1986); National Research Council (1982); Sasson (1984); Sylvester and Kote (1983); U.S. Congress, OTA (1984) and van Hemert, Lelieveld and la Riviere' (1983).
39. Yoxen (1981), pp. 72-73; Buttel and Kenney (1985).
40. Yoxen (1983), p. 228; Dickson (1984), ch. 2.
41. Buttel, Cowan, Kenney and Kloppenberg (1984).
42. Sun (1985), p. 790.
43. Koshland (1985), p. 1191.
44. Busch (1986).
45. Sun (1986), p. 1360.
46. Rogoff and Rawlins, 1985
47. New Scientist (1986), p. 32.
48. Patnaik (1975); Ernst (1980); Goonatilake (1984); Bienfield (1981); Byres (1981); Rose (1985) and Morehouse (1985).
49. Ernst (1981); Goonatilake (1982) and Dickson (1984).
50. Deo, Silva, Busch, Lacy and Mohseni (1986).
51. *News from CGIAR* (1985), p. 3.
52. *Ibid.*
53. Buttel et al. (1984), p. 383.
54. Bhagavantam (1979).
55. *Nature* (1985), p. 65.
56. *Nature* (1985), p. 93; *Economic Political Weekly* (1985), pp. 869-871, pp. 486-

87, pp. 311-12 and Sathyamurthy (1985), pp. 1218-22.
57. Morehouse (1985), and Dembo, Dias and Morehouse (1985).
58. Dutt (1984) and Barros (1985), p. 195.
59. Dorfman (1983), pp. 1, 48; Dickson (1984) and Sun (1985).
60. Hobsbawm (1981), p. 4.
61. Brady (1985), p. 499.
62. Goldstein (1984), pp. 188-189.

References

Abelson, P.H. "Biotechnology: An Overview", *Science*, vol. 219, February 11, 1983, pp. 611-13.

Abrol, D. "American involvement in Indian agricultural research", *Social Scientist*, 11 (10), pp. 8-26.

Aksler, D. "Assembly line plants take root", *Agricultural Research*, vol. 34 no. 4, April 1986, pp. 6-9.

Amin, S. *Unequal Development*, New York, Monthly Review Press, 1976.

Babbage, C. *On the Economy of Machinery and Manufactures*, Fourth Edition (enlarged), London, Charles Knight, Pall Mall East, 1835.

Bagchi, A.K., *The Political Economy of Underdevelopment*, London, Cambridge University Press, 1980.

Balandrin, M., J. Klocke, E. Wurtele and Wm. Bollinger, "Natural Plant Chemicals: Sources of Industrial and Medicinal Materials", *Science*, vol. 228, July 7, 1985, pp. 1154-1160.

Bardhan, P., "Green Revolution and agricultural labourers", *Economic and Political Weekly*, July 1970, Special Number, pp. 1239-46.

Barnet, R.J. *The Crisis of the Corporation*, Institute of Policy Studies, Washington D.C., 1975.

Barnett, P. "Science for Sale: The Pesticide Connection", *Science for the People*, 12 (July/August), pp. 8-35.

Barton, J. "The International Breeder's Rights System and crop plant innovation", *Science*, 216, pp. 1071-1075.

Bhagvantam, S. "Improving Science Education," in *Scientists on Developement: International Scientific and Technological Co-operation Today and Tomorrow*, edited by Colette M. Kinnon, Paris, UNESCO, May 1979.

Bienefeld, M. "Dependency and the Newly Industrializing Countries (NICs): Towards a reappraisal," in *Dependency Theory: A Critical Reassessment* edited by Dudley Seers, London, Frances Pinter, 1981, pp. 79-108.

Bijman, J. "Substitution Processes in European Agriculture," *Trends in Biotechnology*, April 1986, pp. 91-2.

Bloch, E. "Basic Research and Economic Health: The coming challenge", *Science*, vol. 232. May 2, 1986, pp. 595-99.

Booth, D. "Marxism and Development Sociology: Interpreting the Impasse", *World Development*, vol. 13, no. 7, pp. 761-787.

Brady, N.C. "Chemistry and World Food supplies", *Science*, vol. 218, November 26, 1982, pp. 847-53.

Brass, P. "Institutional transfer of technology: The landgrant model and agricultural university at Pantnagar," in *Science, Politics and the Agriculture Revolution in Asia* edited by R.S. Anderson, Paul Brass, E. Levy, and B.M. Morrison, Boulder, CO, Westview Press, 1982.

Brewer, A. *Marxist Theories of Imperialism*, London, Routledge and Kegan Paul, 1980.

Brockway, L. *Science and Colonial Expansion: The Rise of the British Royal Botanical Gardens*, New York, Academic Press, 1979.

Browet, J. "The newly industrializing countries and radical theories of development," *World Development*, vol. 13, no. 7, 1985, pp. 798-803.

Bull, A.T., G. Holt and M.D. Lilly. *Biotechnology: International Trends and Perspectives*. Paris, Organization for Economic Cooperation and Development, 1982.

Busch, L. "Science, Technology, Agriculture and Everyday Life," *Research in Rural Sociology and Development*, vol. 1. 1984, pp. 289-314.

Busch. L. "Can agronomy feed the World? Agricultural research policy and hunger," Paper Presented at The Right to Food: An international Conference on Freedom from Hunger in Montreal, Canada, May 1984, mimeo.

Business Week, "An academic rush for a DNA bonanza," *Business Week*, December 21, 1981, pp. 40-41.

Buttel F.H., J.T. Cowan, M. Kenney and J. Kloppenburg, Jr. "Biotechnology in agriculture: The political economy of agribusiness reorganization and industry-university relationship," in *Research in Rural Sociology and Development* edited by H.K. Schwarzweller, JAI Press, Greenwich, CONN, 1984.

Buttel, F. H. and M. Kenney, "Biotechnology and international development: Prospects for overcoming dependence in the information Age," Cornell Rural Sociology Bulletin Series, Bulletin no. 143, March 1985, Ithaca, NY, Department of Rural Sociology, Cornell University.

Buttel F.H., M. Kenney and J. Kloppenberg, Jr. "Biotechnology and the Third World: Toward a Global Political-economic Perspective," *Politics and the Life Sciences*, vol. 3, 1984, pp. 160-164.

Buttel F. H., M. Kenney and J. Kloppenberg, Jr. "From Green Revolution to Biorevolution: Some observations on the changing technological bases of economic transformation in the third world", *Economic Development and Cultural Change*, 1984.

Buttel F.H., M. Kenney and J. Kloppenberg, Jr. "The IARCs and the development and application of biotechnologies in developing countries," in *Biotechnology in International Agriculture Research*, Manila, International Rice Research Institute, 1984.

Buttel F.H., J. Kloppenberg Jr. M. Kenney and J. T. Cowan. "Genetic Engineering and Restructuring of Agricultrual Research," *The Rural Sociologist*, vol. 3, 1983, pp. 132-144.

Byres, T.J. "The dialectics of India's Green Revolution," *South Asian Review*, January 1972, pp. 99-116.

Byres, T.J. "The new technology, class formation and class action in the Indian countryside," *Journal of Peasant Studies*, July 1981, pp. 405-54.

Caproso, J.A "Industrialization in the periphery: The evolving global division of labour," *International Studies Quarterly*, vol 25, no. 3, pp. 347-384.

Chirot, D. *Social Change in the Twentieth Century*, New York, Harcourt, Brace Jovanovich, 1987.

Clairmonte, F. and J. Cavanagh. *The Transnational Economy: Transnational Corporations and Global Markets*, Institute for Policy Studies, Washington D.C., 1982.

Clairmonte, F. and J. Cavanagh. "Transnational Corporations and Services: The Final Frontier" *Economic and Political Weekly*, vol. 20, no. 8, Feb. 23, 1985, pp. 317-325.

Clairmonte, F. and J. Cavanagh. "Destruction of Sugar Industry," *Economic and Political Weekly*, vol. 21, no. 1, January 4, 1986, pp. 18-19.

Cleaver H. M. Jr. "The Contradictions of the Green Revolution", *Monthly Review*, 24, June 1972, pp. 80-111.

Cooper, C. (Editor) *Science, Technology and Development: The Political Economy of Technical Advance in Underdeveloped Countries*, London, Frank Cass, 1973.

Cooper, I. P. "Patent Protection for New Forms of life," *Federal Bar Journal*, 38, Winter 1979, pp. 34-48.

Comet, R. J. *Les Phares Verts*, Brussels, Editions L. Cuypers, 1965.

Creel, T. and D. Wintringham. "Patent Systems and Their Role in the Technological

Advance of Developing Nations, *"Rutgers Computer & Techonology Law Journal,* vol 10, no. 2, 1984, pp. 255-302.

Dahlberg K. A. *Beyond the Green Revolution: The Ecology and Politics of Global Agricultural Development,* New York, Plenum Press, 1979.

Dasgupta, B. *Agrarian Change and the New Technology in India,* Geneva, UNRISD, 1977.

DeKadt, E. and G. Williams (Editors). *Sociology and Development,* London, Tavistock, 1974.

Delwiche, C.C. "Legumes - past, present and future," *BioScience,* vol. 28, no. 9, September 1978, pp. 565-70.

Dembo, D., C. Dias and W. Morehouse. "Biotechnology and the Third World" Some social economic, political and legal impacts and concerns," *Rutgers Computer and Technology Law Journal,* vol. 11, no. 2, 1985, pp. 431-468.

Dembo, D., C. Dias and W. Morehouse. "The Biorevolution and the Third World," *Third World Affairs* 1985, Third World Foundation for Social and Economic Studies, London, pp. 311-25.

Deo S., J. Silva, W. Lacy, L. Busch and N. Mohseni. "Agricultural Biotechnology in the Third World: Opportunities and Contradictions," Paper presented at the annual meeting of Rural Sociological Society, Salt Lake City, Utah, August 1986.

Deolalikar, A.B. "Foreign Technology in the Indian Pharmaceutical Industry: Its impact on local innovation and social equity," New York, Council on International and Public Affairs, 1980, mimeo.

Dibner, M.D. "Biotechnology in Europe," *Science,* vol. 232, June 13, 1986, pp. 1367-72.

Dickson, D. *The Politics of Alternative Technology,* New York, Universe Books, 1974.

Dickson, D. "Science and Political Hegemony in the 17th Century," *Radical Science Journal,* no. 8, 1979.

Dickson, D. "Conflict of interest on California campus," *Nature,* 293, October 8, 1981, p. 417.

Dickson, D. *The New Politics of Science,* New York, Pantheon Books, 1984.

Dixon, B. *What is Science for?* London, Routledge and Kegan Paul, 1978.

Dixon, B. "The third world's greatest opportunities," *Bio/Technology,* 1:494, p. 519.

Doel, K. van den and G. Junne. "Product substitution through biotechnology: impact on the third world," *Trends in Biotechnology,* April 1986, pp.88-90.

Dogra, B. "Traditional agriculture in India: High yields and no wastes," *The Ecologist,* March 23, 1985, pp. 222-24.

Dogra, B. "Foodgrains exports in a hungry land," *Economic and Political Weekly,* March 23, 1985, pp. 490-91.

Dorfman, P. "International Biotechnology Center declared to exist, but remains without home, funds," *Genetic Engineering News,* November/December 1983, p. 1, 48.

Doyle, J. *Altered Harvest: Agriculture, Genetics and the fate of the World's food supply,* New York Viking Press, 1985.

Dumenil, G., M.Glick and J. Rangel. "Tendency of the rate of profit to fall in the United States," *Contemporary Marxism,* no. 9, Fall 1984, pp. 148-164 and no. 11, Fall 1985, pp. 138-151.

Dumont, R. *Lands Alive,* New York, Monthly Review, 1965.

Dutt, S. *India and the Third World:* Altruism or Hegemony? London, Zed Press, 1983.

Editor, "Unwanted foodgrains," *Economics and political Weekly,* April 13, 1985, pp.

Ernst, D. (Editor). *The New International Division of Labor, Techonology and Underdevelopment,* Frankfurt, Campus Vertag, 1980.

Evans, P. "Continuities and contradictions in the evolution of Brazilian dependence," *Latin American Perspectives,* vol. 8. no. 2, Spring 1976, pp. 30-54.

Evans, P. *Dependent Development,* Princeton NJ, Princeton University Press, 1979.

Faucher, E. "Industrial policy in a dependent state: The case of Brazil," *Latin American*

Perspectives, vol. 12, no. 1, Winter 1980, pp. 117-36.

Feder, E. "McNamara's little Green Revolution, World Bank scheme for self-liquidation of third world peasantry," *Economic and Political Weekly,* vol. 17, no. 14, April 13, 1976, pp. 532-41.

Feder, E. "The Counter-reform bloc," *Ceres.,* March-April 1979, pp. 20-24.

Feder, F. "In Sheep's Clothing," *Ceres* July-August 1976, pp. 40-44.

Folker, F. *The Current Development of the World Economy,* Tokyo, United Nations University, 1980.

Frank, A.G. *On Capitalist Development,* Bombay, Oxford University Press, 1975.

Friedland, W., A. Barton and R. Thomas, *Manufacturing Green Gold: Capital, Labour and Technology in the Lettuce Industry,* New York, Cambridge University Press, 1981.

Friedmann, H. "The Political Economy of Food: The rise and fail of the post-war international Food Order," *American Journal of Sociology,* vol. 88 (Supplement), 1982, pp. S248-86.

Frobel, F., J. Heinrich and O. Kreye. "Tendency toward a New international Division of Labour: Worldwide utilization of labour force for world market oriented manufacture," *Economic and Political Weekly,* Annual Number, February 1976, pp. 159-170.

Frobel, F., J. Heinrich and O. Kreye. *The New International Division of Labour: Strucutral unemployment in industrialized countries and industrialization in developing countries,* Cambridge, Cambridge University Press, 1980.

Fox, J. L. "Biotechnology: A high-stakes industry in flux," *Chemistry and Industry,* March 29, 1982, pp. 10-12.

Galli, A. E. (Editor). *The Political Economy of Rural Development: Peasants, international capital, and the state* Albany NY. State University of New York, 1981.

Galtung, J. "Towards a New International Technological Order," in *Science, Technology and the Social Order,* edited by Ward Morehouse, New Brunswick, NJ, Transaction Books, 1979.

Gandhi, I. "Scientific Endeavour in India, "*Science,* vol. 217, Sept. 10, 1982, pp. 1008-9.

General Accounting Office (U.S.) *Biotechnology: Agriculture's Regulatory System needs clarification,* Report to the Chairman, Committee on Science and Technology, House of Representative, GAO/RCED-86-59, March 1986.

George, S. *Feeding the Few: Corporate control of food,* Institute for Policy Studies, Washington D.C., 19.

Gereffi, G. *The Pharmaceutical Industry and Dependency in the Third World,* Princeton NJ, Princeton University Press, 1983.

Gereffi, G "Power and Dependency in an interdependent World: A guide to understanding the contemporary global crisis," *International Journal of Comparative Sociology,* vol. 25, no. s 1-2, pp. 91-113.

Goldstein, D. J. "The Biotechnological Challenge: To be or not to be like Eduardo S. Bradford," *Interciencia,* vol. 9, no. 4, July-August, 1984, pp. 188-89.

Goodrich, R. S. *An Analysis of Current and Projected Biotechnological Activities* in Brazil, mimeograph, report to the Office of Technology Assessment, Washington D.C., 1982.

Goonatilake, S. *Crippled Minds: An Exploration into Colonial Culture,* New Delhi, Vikas, 1982.

Goonatilake, S. *Aborted Discovery: Science and Creativity in the Third World,* London, Zed Press, 1984.

Griffin, K. *The Political Economy of Agrarian Change* Cambridge MA, Harvard University Press, 1974.

Gutschick, V. P. "Energy and nitrogen fixation," *Bio Science,* vol. 28, No. 9, September 1978, pp. 571-75.

Habermas, J. *Towards a Rational Society: Student Protest, Science, and Politics,* Boston, Beacon Press, 1971.

Hacking, A. J. *Economic Aspects of Biotechnology,* Cambridge University Press, 1986.

Hansen, M. L. Busch, W. Lacy and J. Burkhardt, "Plant Breeding and Biotechnology: New technologies raise important social questions," *Bio Science,* vol. 36, no. 1, January 1986, pp. 29-39.

Harlan, J. "The Green Revolution: Genetic backlash," *Ceres*-The FAO Review, September-October 1969, pp.

Hobsbawm, E.J. *The Age of Capital,* New York, Charles Scribner's, 1975.

Hobsbawm, E.J. "Looking Forward: history and the future", *New Left Review,* 125 (January-February 1981), pp. 3-19.

Hodgkin, L. "Politics and Physical Sciences," *Radical Science Journal,* vol. 4, 1976, pp. 26-60.

Holden, C. "Rockefeller doubles third world efforts," *Science,* vol 232, May 16, 1986, p. 823.

Hurtienne, T. "The Brazilian model of accumulation" Its origins, structure, and crisis," *Latin American Perspectives,* vol. 15, no. s 2-3, Spring/Summer 1983, pp. 108-27.

India. National Biotechnology Board. *Long-range Plan for Biotechnology for India,* New Delhi, Department of Science and Technology, 1983.

Jackson, D. "Biotechnology: Tradition, technology and targets," *Biotechnology in the Americas: Prospects for developing countries,* Washington, D.C., Interciencia Association, 1984.

Jackson, M.W. "Science and Depoliticization," in *Integrated Technology Transfer* edited by Jacques Richardson, Mt. Airy MD. Lamond Publications, 1979, pp. 141-149.

Joyce, C. "Brazil goes back to its roots," *New Scientist,* July 18, 1985, pp. 39-41.

Joyce, C. "Research recoversf from the dark age," *New Scientist,* July 18, 1985, pp. 36-38.

Kay, G. *Development and Underdevelopment,* London, Macmillan, 1975.

Kenney, M., F. H. Buttel, J. T. Cowan and J. Kloppenberg, Jr." Genetic Engineering and Agriculture: Exploring the impacts of biotechnology on industrial structure, university-industry relations, and the social organization of U.S. agriculture," Department of Rural Sociology, Cornell University, Ithaca NY, Cornell Rural Sociology Bulletin no. 125, 1983.

Kenney, M. and F. H. Buttel, "Biotechnology: Prospects and dilemmas for third world development," *Development and Change,* vol. no., 1984, pp.

Khosla, R. K., M. G. Sardana. M. P. Saksena and M. L. Sahni." Review of agricultural experimentation in India during 1965-70," *Indian Journal of Agricultural Science,* vol. 43, no. 4, April 1973, pp. 337-340.

Kowarick, L. and M. Campanario. "Sao Paulo: The price of World city status," *Development and Change,* vol. 17, 1986, pp. 159-74.

Kloppenberg Jr., J. "The social impact of biogenetic technology in agriculture: Past and future," in *The Social Consequences and Challenges of New Agricultural Technologies* edited by G. Berardi and C. Geisler (Editors), Boulder CO. Westview, 1984, pp. 291-321.

Knorr, D. and A. Sinskey. "Biotechnology in food production and processing," *Science,* vol. 229. September 20, 1985, pp. 1224-29.

Koppel, B. "Themes on Genes: Comments on Biotechnology, Agricultural Research, and Rural Sociology," *The Rural Sociologist,* vol. 5, no. 2, pp. 79-88.

Koshland, D.E., Jr. "Excursions in Biotechnology, *Science,* vol. 229 (4719 September 20, 1985., p. 1191.

Laclau, E. "Feudalism and Capitalism in Latin America," *New Left Review,* May-June 1971, pp. 19-38.

Lakoff, S, "Biotechnology and the Developing Countries," *Politics and the Life*

Sciences, vol. 2, no. 2, February 1984, pp. 151-59.

Latin America Weekly Report. "Finance: How much does the region owe?" *Latin America Weekly Report*, January 10, 1986, WR-86-02.

Lehmann, D. (Editor). *Development Theory*, London, Frank Cass, 1979.

Leiss, W. "The false imperatives of technology," in *Thinking about Change*, edited by David Shugarman, Toronto, University of Toronto Press, 1974, pp. 105-21.

Leiss, W. "Technology and instrumental rationality in capitalism and socialism" in Technology and Communist Culture: *The Socio-cultural Impact of Technology under socialism* edited by F. J. Fleron, New York, Praeger, 1977, pp. 115-45.

Leopold, M. "The Transnational Food Companies and Their Strategies," *International Social Science Journal*, vol. 37, no. 3, 1985, pp. 315-330.

Levidow, L. and R. M. Young (Editors). *Science, Technology and the Labour Process*, Atlantic Hts. NJ, Humanities, 1981.

Levins, R. and R. Lewontin. *The Dialectical Biologist*, Cambridge MA, Harvard University Press, 1985.

Lewin, R. "Science and politics in genetic engineering," *New Scientist*, 82, April 12, pp. 114-15.

Lewin, R. "Funds squeezed for international," *Science*, vol. 218, November 26, 1982, pp. 866-67.

Lewontin, R.C."Agricultural research and the penetration of capital," *Science for the People*, vol. 14, no. 1, 1982, pp. 12-17.

Macpherson, C.B. "Democratic Theory: Ontology and Technology" in *Philosphy and Technology* edited by C. Mitcham and R. Mackey, New York, Free Press, 1983, pp. 161-170.

Mooney, P. R. *Seeds of the Earth: A Private or Public resource*, Inter Pares, Ottawa for International Coalition for Development Action, London, 1980.

Mooney, P. R. "The Law of the Seed: Another Development and Plant Genetic Resources," *Development Dialogue* no. s 1-2, 1983.

Moraze, C. Science and the Factors of Inequality: *Lessons of the Past and Hopes for the Future*, Paris, UNESCO, 1979.

Morehouse, W. *Science, Technology and Social Order*, New Brunswick NJ, Transaction Books, 1979.

Morehouse, W. "The New Levers of Global Power: Science and technology in the international political economy," *Third World Affairs 1985*, Third World Foundation for Social and Economic Studies, London, 1985, pp. 291-300.

Morgan, D. *Merchants of Grain*, New York, Viking, 1979.

Moseman, A. *Building Agricultural Research Systems in Developing Nations*, New York, Agricultural Development Council, 1970.

Nakaus, H. "Japan's internationalization: Becoming a global citizen" *International journal of Comparative Sociology*, vol. 25, no. s 1-2, 1984, pp. 114-122.

National Commission on Agriculture. *Report on Research, Education, and Extension*, New Delhi, Government of India, 1976.

National Research Council. *Priorities in Biotechnology Research for international Development*, Washington D.C. National Academy Press, 1982.

New Scientist. "Foreign fields save western crops," *New Scientist* October 28, 1982, p. 218.

Nielson, K. "Techonology as ideology" in *Research in Philosophy and Technology* edited by Paul Durbin, vol. 1, Greenwich CONN, JAI Press, 1978, pp. 131-147.

Noble, D. F. "The selling of the university," *The Nation*, February 6, 1982, pp. 143-48.

Norman, C. *The God that Limps*, New York, W.W. Norton, 1981.

Oasa, E. "The political economy of international agricultural research: A review of the CGIAR's response to criticisms of the 'Green Revolution'." Unpublished manuscript.

Oasa, E. and L. Swanson. "The Limits of Farming Systems Research and Development: Should development administrators be interested?", *Human Organization*, 1986.

O'Connor, J. *The Fiscal Crisis of the State*, New York, St. Martin's Press, 1973.

Offe, C. "The Theory of the Capitalist State and the Problem of Policy formation," in *Stress and Contradictions in Modern Capitalism* edited by Leon Lindberg, Robert Alford, Colin Crouch and Claus Offe, Lexington, MA, D.C. Health, 1972.

Ozawa, Y. "Common Problems in Tissue Culture Work in Developing Countries," in *Practical Tissue Culture Applications* edited by Karl Navamorosch and Hirumi Hiroyuki, New York, Academic Press, 1979, pp. 45-56.

Palma, G. "Dependency" A Formal Theory of Underdevelopment or a Methodology for the analysis of Concrete situations of Underdevelopment?" *World Development*, vol. 6, 1978, pp. 881-924.

Pearse, A. *Seeds of Plenty, Seeds of Want* New York, Oxford University Press, 1980.

Ping, H.K. "Revolt of the landless peasants." *Far Eastern Economic Review*, January 12, 1979, pp. 59-60.

Plucknett, D. L. and N. J. H. Smith, "Agricultural Research and Third World Food Productions," *Science*, 217, July 16, 1982, pp. 215-220.

Reddy, A.K.N. "The Trojan Horse." *Ceres*, March-April 1976, pp. 40-43.

Rehman, A. Triveni: *Science, Technology, and Socialism* Simla, Indian Institute of Advanced Study, 1977.

Rehman, A. "The interaction between science, technology and society: historical and comparative perspectives", *International Social Science Journal*, vol. 33, no. 3, 1981, pp, 508-21.

Reichert, W. "Agriculture's Diminishing Diversity: Increasing yields and vulnerability." *Environment*, vol. 24, no. 9, November 1982, pp. 6-11, 39-43.

Richardson, J. *Integrated Techonology Transfer*, Mt. Airy, M. D. Lamond Publications, 1979.

Rogers, M.D. "The Japanese Govt.'s Role in Biotechnology R&D," *Chemistry and Industry*, August 7, 1982, pp. 533-37.

Rogoff, M. and S. Rawlins. "Food Security: A Technological Alternative," mimeo, June 1985.

Rose, D.A. "Changing Relationships in Technology Transfer to the Third World: Case Study of Biotechnology Agriculture," *Rutgers Computer and Technology Law Journal*, vol. 11, no. 2, 1985, pp. 531-557.

Rosenberg, N. "Karl Marx on the Economic Role of Science." *Journal of Political Economy*, vol. 82, no. 4, July-August 1974, pp. 713-28.

Safeeulla, K. M. "Genetic vulnerability: The basis of recent epidemics in India." *Annals of the New York Academy of Sciences*, 287, pp. 72-85.

Sardar, Z. and D. G. Rosser-Owen, "Science Policy and Developing Countries," in *Science, Technology and society: A Cross-Disciplinary Perspective* edited by ina Spiegel-Rosing and Derek de Solla Price, London, Sage, 1977, pp. 535-75.

Sasson, A. *Biotechnologies: Challenges and Promises*, Paris, UNESCO, 1984.

Schmidt, A. *The Concept of Nature in Marx*, London, New Left Books, 1971.

Schroeder-Gudehus, B. 'Science, Technology and Foreign Policy" in *Science, Technology and Society: A Cross-Disciplinary Perspective* edited by Ina Spiegel-Rosing and Derek de Solla Price, London, Sage, 1977, pp. 473-506.

Schaffer, B. and Wen-hsein, H. "Distribution and the theory of access." *Development and Change*, vol VI, no. 2, 1975, pp. 13-36.

da Silva, J.F., Graziano, D. A. and Kohl, B. A. "Capitalist modernization' and employment in Brazilian agriculture, 1960-1975: The case of the state of Sao/Paulo." *Latin American Perspectives, II* (1), Winter 1984, pp. 117-36.

Smith D.H. and J. King. "The Legislative and Legal Background (to PVPA)," *Environment*, vol. 24, no. 6, July-August 1982, pp.24-26.

Sohn-Rethel, A. "Science as Alienated Consciousness," *Radcial Science Journal*, vol. 2, no. 3, pp. 65-101.

Spiegel-Rosing, I. and D. de Solla Price (Editors). *Science, Technology and society: A Cross-disciplinary Perspective*, London, Sage, 1977.

Stanley, D. W. "Chemical and Structural Determinants of texture of fabricated foods," *Food Technology*, March 1986, pp. 65-76.

Sun, M. "The Japanese Challenge in Biotechnology," *Science*, vol. 230, November 15, 1985, pp. 790-92.

Sun, M. "Engineering Crops to resist weed killers" *Science*, vol. 231, March 21, 1986, pp. 1360-61.

Swaminathan, M. S. "Biotechnology Research and Third World Agriculture," *Science*, vol. 218, December 3, 1982, pp. 967-72.

Sylvester, E. J. and L. C. Klotz. *The Gene Age: Genetic Engineering and the Next Industrial Revolution*, New York, Charles Scribner's Sons, 1983.

Tweeten, L. *High Technology and Rural America*, Stillwater. OK. Oklahama State University, Department of Agricultural Economics, prepared for the office of High Technology Education at the University of Tennessee, unpublished manuscript.

U.S. Congress, Office of Technology Assessment (OTA). *Commercial Biotechnology" An International Analysis*, OTA, Congress United States, Washington D.C., OTA-BA-218, January 1984.

U.S. Congress, Office of Technology Assessment (OTA). *Technology, Public Policy, and the Changing Structure of American Agriculture*, OTA-F-285, Washington D.C., U.S. Government Printing Office, March 1986.

Vandermeer, J. "Tomatoes in the Midwest" Agricultural Research and Social Conflict," *Science for the People*, January-February 1981, 5-8, pp. 25-30.

Vandermeer, J. "Science and Class Conflict: The Role of Agricultrual Research in the Midwestern Tomato Industry" in *Studies in Labour Theory and Practice, Studies in Marxism*, vol. 12, edited by William L. Rowe, Minneapolis, Marxist Educational Press, 1982, pp. 41-58.

Van Hemert, P.A., H.L. M. Lelieveld and J.W.M. la Riviere *Biotechnology in Developing Countries*, Delft, the Netherlands, 1983.

Vergapoulas, K. "The end of agribusiness and the emergence of biotechnology," *International Social Science Journal*, vol. 37, no. 3, 1985, pp. 286-99.

Wade, N. "Green Revolution (I): A just technology, often unjust in use," *Science*, 186, December 20, 1974, pp. 1093-96.

Wade, N. "Green Revolution (II): Problems of adapting a Western Technology," *Science*, 186, December 27, 1974, pp. 1186-92.

Walsh, J. "Germplasm resources are losing ground," *Science*, 214, October 23, 1981, pp. 421-23.

Westergaard, J. and M. Ressler. *Class in a Capitalist Society*, Harmondsworth, Penguin, 1976.

Wilkes, G. "The Endangered Genetic Base of the World's Food Supply," *The Bulletin of the Atomic Scientists*, February 1977, pp. 8-16.

Wolf, E.C. "Challenges and Priorities in Conserving Biological Diversity," *Interciencia*, vol. 10, no. 5, September-October 1985, pp. 236-242.

Yoxen, E.J. "Life as a Productive Force: Capitalizing the Science and Technology of Molecular Biology" in *Science, Technology and the Labour Process* edited by L. Levidow and R. Young, Atlantic Highlands NJ, Humanities, 1981.

Yoxen, E.J. *Gene Business: Who Should Control Biotechnology?* London and Sydney, Pan Books, 1983. Yoxen, E.J. "The Social Impact of Biotechnology," *Trends in Biotechnology* , April 1986, pp. 86-88.

Zilinskas, R.A. "New Biotechnology: Potential Problems Likely Promises," *Politics and the Life Sciences*, vol. 2, no. 1, August 1983, pp. 42-52.

3

LAND TO LAB: Some Issues in Technology Creation in Modern Agriculture

C.R. PRASAD RAO & SUSAN VERGHESE

If one were contemplating the construction of a theory of technology creation in agriculture, consideration would embrace a wide variety of factors with varying relative emphases, such as (1) relative factor endowments and scarcities (e.g. land, labour, capital, and water); (2) export orientation versus domestic consumption; (3) vested interests of farming populations and their political clout, and organized lobbying; (4) market distortions and factor and product prices; (5) fads, biases, and predilections of policy makers and research managers; (6) existing state of technology and scientific manpower; (7) science-related overheads and infrastructure; and (8) professional and personal orientations, commitments, and interests of agricultural scientists. Thus, one can come up with a largely economic theory of technology creation, or a purely managerial and human relations theory, or a political economy theory, or a social and psychological theory, depending upon the relative importance accorded to the various factors. A sociology of knowledge framework, which treats technological change as a special variant of social change, seeks to offer a comprehensive, synthetic, holistic perspective attempting to integrate the various sets of determinants of technological change. In this framework the agricultural scientist, the science and technology he creates, and his professional behaviour, values, and interests are all socially situated. In a framework of this kind, it is not safe or valid to assume that agricultural technologies created are entirely or strictly in conformity with the perceptions and estimations of user needs, conditions, and priorities. Particularly in a Third World country like India, characterized by skewed land distribution, preponderance of small and marginal operators, adverse man-land ratios, and subsistence farming systems, the technology creation process would rarely be free of latent and manifest distortions. This situation prevails for the simple reason that marginalized subsistence sectors of Indian agriculture, scattered all over, and indigent and unorganized, have not yet constituted powerful pressure groups to articulate and bargain for their interests with the architects of technology policies, technology makers, and research administrators. There is as yet no impressive evidence of direct participation of this section of the agricultural community in the technology creation processes. Their needs, problems, and interests are communicated indirectly, if at all, by their so-called spokesmen – the field extension personnel, and to a large extent by the members of the fourth estate, journalists and agricultural correspondents. For all practical purposes, no viable, assertive and articulate cooperative movement is emerging, and the political parties that espouse their interests are weaklings on the Indian political scene. The glaring case of the dryland agricultural sector in India substantiates this contention. Although this sector is predominant in terms of area and population, it received until now only five percent of the total research budgets of the Indian Council of Agricultural Research (ICAR). It is only in the last few years or so this sector is being accorded the attention it deserves in research and policy making circles and, even so, a reasonable surmise is that the dry sector is attracting notice for no other reason than the short-fall in pulses and oilseeds production and the consequent high prices and import bills.

Commenting on the lacunae in farm research, Dr. A. Appa Rao, Vice-Chancellor of the Andhra Pradesh Agricultural University observes (1985: 14):

> "Researchers have not been successful in providing the relevant technology to support increase in production, especially in limited-resource, risk-prone situations. One reason is inadequate pressure from the farming community for new appropriate technology. Another is the limited field exposure of researchers. A third is the drawback in research programme priorities and formulations–top-down approach in research planning.... The field trials conducted are done more to fulfil the requirements for varietal release than to obtain comprehensive information on the on-farm performance. Researchers tend to shun tough risks. Farmers adopt practices which provide maximum management and operational feasibility and economic stability. Knowing about the farmer, let alone from

> him, has yet to be accepted at least in practice as part of the professional work of an agricultural researcher... our research is discipline-oriented instead of being more farmer and problem-oriented."

Another commentator has no hesitation to report that the people behind the lofty claims for India's Green Revolution are mostly the agricultural scientists and technocrats themselves.[1]

Whatever be the truth or exaggeration in this allegation, it calls attention to the important issue of the degree to which scientific temper and culture prevails among agricultural scientists. If scientific temper implies an open, unbiased, objective frame of mind in which to examine every critical issue of national importance, has it emerged as a cultural trait, as a basis of research activities and priorities? Is scientific rationality a mightier force than primordialism and vested interest?[2] Will not the thin veneer of scientific rationality melt away at the allurements of positions of power, media publicity, and international scientific rewards?[3,4,5] As the great dedicated technologist, Seshadri (1983), observes: if science should echo cultural values, it should have a grassroots basis and orientation, a relevance to life and environment in which it evolves or is applied. The true measure of a science is not C.V. Raman winning a Noble Prize or S.N. Bose co-authoring a paper with Einstein. The ethos and assumptions of science must reflect the priorities of society. The notion of catching up and the resulting sophistication and internationalism in science and technology creation are unhelpful to this goal. Instead, the direct involvement of the scientist in work in rural areas helps a great deal. Science has no readymade answers for every local problem (William and Paul Paddock, 1971: 718). Grassroots exposure leads to not only local adaptation but paradoxically to even the transformation of science and technology.[6]

Features of Agriculture

In contrast to other productive enterprises, agriculture presents a few outstanding features. Zimmerman's is an excellent account of these features (1971: 17-29). The most prominent of them is the agro-climatic heterogeneity. Agriculture is a techno-economic activity characterized by immense diversity of farming situations all over the world. The diversity owes mainly to three factors: its anciency; its dependency on nature; and its global diffusion. Dependent on nature, agriculture's diversity parallels nature's diversity. As the art and craft of producing food, fuel, fodder, and fibre, agriculture is caught up in a complex ecosystem – water-soil-plant-seed-climate-insects and micro-organisms – of nature. Agricultural technology concerns the use and management of this factor matrix and their interactive relationships. As Zimmerman writes, agriculture concedes to nature the dominant role in man's productive efforts. In it man has not gained the upper hand; nature remains the master and man is only the helper (1971: 18-19). Man's managerial ability and control decreases progressively from, say, water and soil, to insects and micro-organisms and the biotic energies of plants and animals, to climate (temperature, rain, wind and solar-radiation). Modern agriculture may increasingly come under the spell of machines, science, industry, and commerce, but it cannot be wholly absorbed by them, according to Zimmerman. Agriculture retains an irreducible residual dependence on living nature (1971: 25). This is best illustrated by plant and animal responses at various levels – cellular and intracellular, biochemical, physiological and yield levels – to pollutants and agro-chemicals (see the companion article).

A second critical feature of agriculture is that cultural and socio-economic heterogeneity overlays its natural diversity. The direct consequence of anciency and the global diffusion of agriculture is that it is practised not only under every conceivable combination of natural factors but also under a vast diversity of socio-economic conditions and cultural institutions. The cultural and socio-economic heterogeneity interlocks with the ecosystem diversity in agriculture. Technological possibilities and outcomes are constrained by this overlay of heterogeneities of farming systems.

The third outstanding feature of agriculture is the highly privatized character of the production decisions, skills, and behaviours of millions of farm producers. Naturally, technology creation, technology acceptance, and technology diffusion are affected and complicated by these features of agriculture, whereas the natural diversities preclude possibilities of creation of universal agricultural technologies and impose the harsh condition that a workable technology must necessarily be crop- and locality-specific. The agronomic and socio-economic heterogeneity, and individual farm decision-making restrict the adoption and diffusion of specific kinds of technology packages. A high-cost, high-input, high-risk, low-profit, and low-product-quality technology is bound to diffuse falteringly and selectively (Prasad Rao and Chakrapani, 1985: 1-29). Agro-climatic and socio-economic heterogeneity and privatized decisions together prodcuce a situation in which it may be safely hypothesized that where field extension agency networks are more effectively established and organized in servicing the needs of technology producers and technology users, there is a greater chance for diverse and appropriate technologies to be forged from the laboratories because of more efficient land-to-lab feedback. The three features of agriculture discussed suggest the need for building a strong organizational base of triadic relations among farmers, agricultural scientists and extension personnel. Isolation of farm scientists from farmers and farming situations exists in a mediated organizational situation as indicated below:

Scientists ⇄ Extension Agents ⇄ Farmers

In a alternative organization of the information system the relationship between scientists and farmers is direct and unmediated as indicated in the triadic model below:

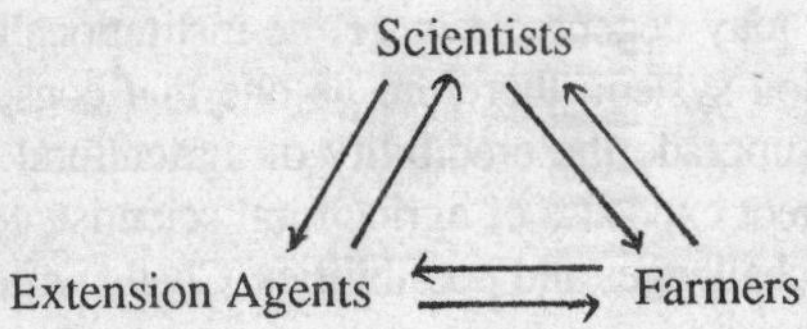

In the circular triadic model of organization farm research acquires more relevance, locality-specificity, and problem-centredness. Agricultural research tends to become more dynamic, diverse, and innovative because of grassroots planning. It helps overcome homophily in communicative behaviour and enables the scientist to assess and define the user needs more accurately and continuously in terms of required research operations, reorientations, and adaptations. In the indirect, mediated two-step communication system farmers as well as the extension personnel may not be able to identity and define the problem accurately to the scientist. Furthermore, the alternative system helps the scientist acquire a holistic picture of the problem situation. Modernization of agriculture, thus, depends as heavily upon effectively meeting the field informational needs of farm scientists as it depends upon lab-to-land information transfer. This, rather than what Lionberger and

Chang suggest, is the crux of the communication problem of technological development in agriculture. Lionberger and Chang write:

> "The major communication problem in technological development... is essentially worldwide diffusion of ideas and concepts, not independent invention" (1970: P.V).

Sustained, direct interaction with ultimate users of technology and local adaptive research provide effective feedback to agricultural scientists. Extension personnel alone should not be entrusted with providing land-to-lab feedback and local adaptive testing. An integrated system of information management rather than a differentiated system seems to fit the situation well (Lionberger and Chang, 1970:4). An endemic danger of the functionally differentiated two-step farm-information model is the tendency for a gradual and progressive isolation of agricultural scientists from farmers and farming situations and for their complete dependence on extension agencies for land-to-lab feedback, and an unconscious tendency to overrate research roles and underrate extension functions. In actual practice, more emphasis may come to be laid on lab-to-land transfer of information and the land-to-lab system of information may degenerate or get de-institutionalized. An efficient farm information system, therefore, is one that constantly endeavours to sustain and upgrade the credibility of agricultural research by continuous and direct exposure of agricultural scientists to farm-originated stimuli, ideas, challenges and possibilities. Chang writes:

> "Both extension personnel and researchers will have to travel all over the country and look for problems which are limiting factors in crop production. Though many of these problems can be solved with little effort, the impact is often very far-reaching. With the solution of some of their pressing problems, the farmer's confidence in the extension and research workers is built up and they become more receptive to new ideas and new techniques" (1971: 213).

Citing various reasons for non-adoption of improved dryland technology (e.g. risk and uncertainty, pricing policy, and soil-climatic variation), dryland agriculture researchers recommend that

> "more emphasis is to be laid on the development of graded technol-

ogy treating various components of technology in isolation as well as in combination, rather than developing a full package of technology for general adoption." (Reddy, Rastogi, Rao and Korwar, 1985: 1).

The common experience of those dealing with agricultural development is that agricultural scientists have a tendency to take the farmer for granted. The assumptions underlying their research planning and priorities may be uninformed or even misinformed. Scientists are reluctant to work at the hardest end of the research spectrum where quick results are not possible. They have a tendency to develop vested interest in a specific research scheme affecting their security, survival and career prospects. There was no broad-based development of agricultural research to give technology support proportionately to all areas and all important crops.

Weak Research Base and Emphasis

The large-scale sustained adoption of a technology package depends ultimately on the year-to-year stability of crop yields and the economic value of the whole package. Therefore, stability, productivity and returns-to-costs considerations assume importance in the evaluation of agricultural research. Two important questions concerning a nation's agricultural research agenda and organization are resource allocations for agricultural research and its productivity. Since allocation for agricultural research forms part of overall scientific research outlays, the following data were gathered from sources readily available. Treating science and technology manpower as a productive and manpower base, India's relative emphasis on and investment in research is very low in an international comparison. The following ratios of scientists per thousand population illustrate this:

Country		Scientists per 1000 population
	(1978-79 figures)	
India	:	3
USA	:	15
USSR	:	90
Japan	:	180

Source: *Science Today*, January, 1983, p. 9.

Central Government	:	138,331
State Government	:	22,490
Private Sector	:	23,214

Source: G.S. Aurora, 1986: 1.

A large part of this manpower is not actually engaged in activities that may be regarded as truly scientific and technological. Its quality also varies widely.[7] A large percentage of science and technology workers are in the central sector as the following data for the year 1980-81 show:

The break-up of the total Sixth Plan outlay for Science and Technology research by Agency is furnished in the following table:

Table 3.1: **Sixth Plan outlay for Science and Technology Research by Agency:**

Agency	Plan outlay (Rs. in crores or 10 millions)	Non-Plan outlay (Rs.in crores or 10 millions)
Atomic Energy	248.98	284.59
Space	245.80	146.92
Science and Technology	134.87	189.55
Environment	40.00	13.12
CSIR (Council of Scientific and Industrial Research)	170.00	218.24
ICAR (Indian Council of Agricultural Research)	340.00	190.00
ICMR (Indian Council of Medical Research)	40.00	26.00

Source: *Science Today*, April, 1981, P. 29.

Refflecting on science and technology in the Sixth Plan, Haksar, a retired Deputy Chairman of the Planning Commission, laments that the imperatives of politics and the imperatives of science and technology hardly coincide except in rhetoric. Large areas of economic activity are performed with relatively obsolete and cost-ineffective technologies and the pace of scientific and technological innovation is unimpressive and tardy (1981:11-13).

A picture of the yearly pattern of expenditure on science and technology for the three sectors is available from Table: 3.2

Table 3.2: **Expenditure on Science and Technology**

Rupees in crores or 10 millions

Sector	1974-75	1975-76	Year 1976-77	1980-81	1981-82	1982-83
A. Central Government	258.92	323.70	361.49	617.22	765.15	1108.1
B. State Government	28.65	31.94	37.14	75.73	91.00	126.13
C. Private Sector	36.46	42.35	49.50	120.69	147.00	193.65
Total:	324.04	397.99	448.13	813.64	1003.15	1428.87

Source: *India*, 1980, p. 74
India, 1984, 86, Government of India, Ministry of Information and Broadcasting Publications.

From Table 3.2, it is noted that the total expenditure on science and technology has increased from Rs. 324.04 crores in 1974-75 to Rs. 1,427.87 crores in 1982-83. This represents a little more than four-fold increase over a period of six years. In comparison, the expenditure on Science and Technology in the private sector has increased from Rs. 36.46 crores in 1974-75 to Rs. 193.65 crores in 1982-83, a more than five-fold increase. There is relatively a slower tempo of growth in Science and Technology expenditure in the central sector than in the private and state sectors. The Science and Technology expenditures per worker in the three sectors for the year 1980-81 are Rs. 44,619.07 in the Central, Rs. 33,672.74 in the State, and Rs. 51,990.18 in the private sectors. In other words, scientists and technologists employed in the private sector have higher per-capita expenditures than their counterparts in the other two sectors and the differences are considerable. Thus, two striking findings from the data are that science and technology research shows a quicker growth trend and a higher per worker expenditure in the private sector than in the central and state sectors.

Presuming that little agricultural research gets done in the private sector, the better trends seen for the private sector indicate an investment bias in this sector towards industry. The relative position of agricultural research vis-a-vis other kinds of research in the central sector is revealed in the data presented in Table 3.3.

Table 3.3: **Budget for Major Scientific Agencies in the Central Sector**

(Rupees in Lakhs)

	Agency	1977-78	1980-81	1981-82	1982-83
1.	Atomic Energy	5,706.00 (18.87)	7,347.81 (16.76)	8,826.06 (16.13)	10,563.13 (15.10)
2.	Council of Scientific and Industrial Research	4,667.00 (15.43)	6,900.00 (15.74)	7,877.08 (14.40)	10,081.50 (14.42)
3.	Defence Research	5,145.00 (17.02)	7,970.00 (18.18)	10,483.36 (19.16)	12,199.96 (17.44)
4.	Indian Council of Agricultural Research	5,918.00 (19.58)	9,744.67 (22.23)	11,149.50 (20.38)	13,121.50 (18.76)
5.	Indian council of Medical Research	449.00 (1.49)	900.11 (2.05)	1,195.13 (2.18)	1,423.15 (2.03)
6.	Department of science and Technology	3,991.00 (13.20)	4,063.69 (9.27)	4,608.01 (8.42)	6,051.06 (8.65)
7.	Space Research	3,839.00 (12.70)	5,601.56 (12.78)	7,503.44 (13.71)	8,605.17 (12.30)
8.	Electronics	517.00 (1.71)	540.55 (1.23)	751.11 (1.37)	857.55 (1.21)
9.	Environment		373.99 (0.85)	546.88 (1.00)	905.07 (1.29)
10.	Ocean Development			788.43 (1.44)	4,935.94 (7.06)
11.	Non-conventional Energy Sources		400.44 (0.91)	989.82 (1.81)	1,203.20 (1.72)
	Total	30,232.00 (100.00)	43,842.82 (100.00)	54,718.82 (100.00)	69,947.23 (100.00)

Note: Figures in parentheses indicate percentages.
Source: *India*, 1980, p. 74
India, 1984, p. 86, Government of India Publication, Ministry of Information and Broadcasting.

The budget for agricultural research (i.e. ICAR) was 19.58 percent of the total in 1977-78, 22.23 percent in 1980-81, 20.38 percent in 1981-82 and 18.76 percent in 1982-83. In fact, the allocations for this agency were consistently higher in all the years considered. Defence Research and Atomic Energy came close to the ICAR in the budget allotments. Thus, we see that in the central sector budget allocations, agriculture receives a higher share, although it generally hovered close to one-fifth of the total allotments. The per-worker budget data for five major scientific agencies in the central sector are worked out in Table 3.4

Table 3.4: **Per Capita Budget Allocation for Major Scientific Agencies in the Central Sector for the Year 1980-81**

	Agency	Number of workers engaged	Budget allocation (rupees in lakhs)	Per worker budget in rupees
1.	Atomic Energy	16,402	7,347.81	44,798.26
2.	Council of Scientific Industrial Research	20,303	6,900.00	33,985.13
3.	Defence Research	21,306	7,970.00	37,407.30
4.	Indian Council of Agricultural Research	17,884	9,744.67	54,488.20
5.	Indian Council of Medical Research	2,156	900.11	41,749.07

The data presented in Table 3.4 support the observation that ICAR receives a relatively higher allotment in the central sector. The per-worker budget for ICAR in 1980-81 was Rs. 54,488.20 which is higher than that of Atomic Energy (Rs. 44,798.26), Medical Research (Rs. 41,749.07), Defence Research (Rs. 37,407.30) and Industrial Research (Rs. 33,985.13).

Multiple Research Goals

Attempts have been made, especially by economists, to measure returns to investment in agricultural research, cropwise and countrywise for specified periods, since it is generally believed that agricultural research represented a relatively better investment opportunity than other developmental investments (Arndt and Ruttan, 1975: 1-19). While several of the estimates of returns to agricultural research are questionable on grounds of accuracy, methodology and subjectivity, it is argued that, as quantitative measures, they are nevertheless useful to monitor performance of agricultural research (Arndt and Ruttan, 1975: 5-6). From the returns point of view, three ways by which agricultural research contributes are (a) in finding out ways and means of reducing production costs, (b) in determining the combination of practices and inputs that increase yields, and (c) in developing the most efficient post-harvest technologies for processing and storage, preservation, and marketing of products to enable the farmers to reap the benefits of higher prices. Agricultural research may also be directed to increasing labour employment, monitoring changes in natural systems in consequence of farming, changes in soils, nutrient cycles, and flora and fauna in systems subjected for long to traditional and modern farming, and tolerance parameters of agroecosystems. Yet another fruitful avenue of research is the exploration of land-use systems and their impact on health, ecology and socio-economic conditions of the local populations (Hecht, 1982). The focus on economic returns to research costs appears a narrow concern viewed in the light of this broad spectrum of objectives of agricultural research.

Making a critical note of India's agricultural experience, Hanumantha Rao (1986: 229-233) observes that the present yield-increasing, land-saving agricultural technology is also labour-saving, and, hence, not suitable for large tracts of the country characterized by low investment and risk-bearing capacity. What is needed is a technology which raises the total output per unit of land from all crops through multiple cropping, augmenting cropping intensity, and labour absorption. "The scientific infrastructure borrowed by the developing economies is conditioned by the factor-use biases of the developed economies." (Hanumantha Rao, 1986:230). Zimmerman dichotomizes agricultural research as "true" or assets—preserving and non-polluting, and "spurious" or destructive and pollutive (1971: 22-26). There are multiple ranges and layers of consequences of technological change, all of which are not comprehensible and predictable in advance. As a matter of strategy, technology creation and technology evaluation must go on concurrently to avert irreversible dangers of technological innovations. The present seed-fertilizer-pesticide technology is admittedly high-yielding, but is also high-cost, high-risk and high-pollution, and quite incompatible with the socio-economic and eco-agrarian conditions of large numbers of primary producers. Its inequitable distributional, and disastrous ecological consequences are increasingly being recognized and discussed.

For Third World Countries, the most appropriate agricultural technology is one that takes into account the energy, infrastructural, ecosystemic and socio-economic constraints of the numerically preponderant sections of the farming community (Hecht: 1982; Whyte and Beynton: 1984; Berardi and Geisler: 1984; and Mooney and Gordon: 1983).

Two major issues of modern agricultural technology, in general, merit special attention. These concern (a) yield gaps and yield variations and (b) stability of adoption behaviour of farmers. Experience with improved dryland agricultural technology shows considerable slippages in yields from research farms to farmers' fields, from verification trials on farmers' fields to demonstration farms, from demonstration farms to progressive farmers' fields, and between ordinary farms (Reddy, Rastogi, Rao and Korwar, 1985: 3-9). Such yield gaps, induced by climate, cultural and managerial practices, soil conditions, and farm size and technology, are observed between different plots of the same farmer, and between different farms within the same agro-climatic zone. Graded and variable technologies rather than uniform technology packages are an answer to this problem. The development of graded and variable technologies requires direct exposure of agricultural scientists to farmers' situations rather than a strong land-to-lab communication system. The reason is, studies show evidence of differences in need and priority perceptions between farmers and extension workers (Singh, 1976 and Haque and Singh, 1972: 235-241). The modern Indian farmer is becoming technologically discriminative, and more sophisticated and exacting in his informational needs which are precise and specialized. The questions he asks are more pragmatic in terms of the peculiar individual problems he confronts in his farming situation (Mishra, 1977:10). The possibilities of developing graded and variable technologies are realized much better with direct and continuous exposure of agricultural scientists to farmers and their farming problems than with mere reliance on local, adaptive research, or linear land-to-lab communication systems. Agricultural research must be oriented to attainment of more uniform yields around the best-practice package and reduce divergencies in yields between the best-practice farmers and the farming majority. In this sense, then narrowing of variations in yields among important segments of farmers would be as important as aiming to increase yields of the best farmers and in the best regions (Joshi and Kaneda, 1982: A1-A8).

The second issue of modern agricultural technology refers to role-switching behaviour of farmers who may be classified as total adopters, partial adopters, non-adopters and so on. A total adopter is one who adopts the entire package at recommended levels. He may not commit all his land to the new technology. He is like the adopter at the trial stage of adoption. A partial adopter accepts only a few of the recommended practices or uses them at adjusted or modified levels. He may again be a full-scale or limited-scale partial adopter. A total or partial adopter may, in course of time, turn into a non-adopter by discontinuation. Thus, adopters, partial-adopters and non-adopters are not fixed categories or stable roles. This role-switching behaviour of farmers provides a crucial test of the success and failure of technology. Success involves conversion from partial and non-adoption to total, full-scale adoption of the largest number of potential adopters. A conversion experience in the reverse order is disastrous to extension work. The change-over from adoption to non-adoption calls into question the validity, viability, stability and profitability of the recommended technology. It also undermines the credibility of agricultural research and extension. Therefore, the trialability and compatibility of an innovation and its associated post-harvest problems including prices should be carefully assessed by teams of farm scientists, economists and sociologists. A diachronic, multifaceted framework of investigation of agricultural technologies will be called for. No technological breakthrough in agriculture can become revolutionary unless role-switching takes place in the desired direction on a large scale and sustained basis. A report on "Integrated Dryland Agricultural Development Programme in Ibrahimpatnam Taluk in the State of Andhra Pradesh, South India" (Agro-Economic Research Centre, 1981) furnishes ample evidence of the role-switching behaviour of adopters. This was a study first conducted on 93 farmers in 1972-73 and repeat study was made on 76 of the farmer 93 respondents in 1978-79. The rejection rate in both the studies was about 25 percent. The rejection rate was generally higher among large farmers (60 percent) than small farmers (40 percent). Considering the data for the total adopters who adopted all the recommended practices, considerable slippage was noticed between the two studies. While 30 farmers out of 93 have reported using all the recommended practices in the 1972-73 study, only two respondents out of 76 in the 1978-79 repeat study have reported continued use of all the recommended practices. The discontinuation rate was quite steep over a six-year period. Total adopters became partial adopters in large numbers. Such conversion experiences cast doubts on the quality and relevance of the technology sought to be extended.

New Technology Implications

The kind of technology developed and put through the delivery system currently for use by farmers is a biochemical seed-fertilizer-pesticide package. This technological package is just one coordinate of the production system. The other coordinates are the farmers; the hierarchical socio-economic structure of the agrarian organization; the ecological and the geo-climatic factor; the factor and product markets; the public policy (subsidies, price supports, levy and public procurement and distribution, internal movement of produce and external exports, etc.); and lastly the social overheads (extension service, storage, power, roads, transport and marketing). The adoption of this technology meant two dislocative events for farmers: replacement of traditional farming practices and incorporation of primary producers into a wider network of relations and dependency. In the earlier farming systems, the farm input requirements were met by the farmer from his own farm and village resources. This feature was dissolved and the farmer is drawn out of the circle of his traditional relationships and institutions into a new and more complicated nexus of relationships and institutional arrangements. The traditional farmer could raise the needed seed on his own farm or procure it from neighbours. In the new technology, the dependency on outside agencies created a situation of uncertainty with untimely or inadequate supply or poor quality of seed affecting the cropping schedule and performance. It made him vulnerable to exploitation (e.g. adulterated, spurious and diseased or unhealthy seed). This is because the high yielding varieties are not genetically stable and cannot be bred by farmers. For example, in rice it is difficult to develop a variety that has a useful life for more than five to six years in tropical environment unless genes for more stable resistance are identified and incorporated (Swaminathan, 1983). For instance, in the I.R.-20 variety of rice the hybrid quality gradually fades away after two years of successive cultivation with the same seed.[8] Seed multiplication and distribution is another problem associated with hybrid varieties. Malpractices in seed certification agencies were reported.[9] The problem facing the seed multiplication agencies is the difficulty of assessment of seed requirements of farmers in advance, creating delays and over on undersupplies.[10] In the traditional system, the farmers were growing their own seed at no extra cost and raising the required seed was built into the farm routine. They had full control over the quantum and quality of the seed. From years of experience, they could easily identify which seed material was good and which was not, and they knew how to prepare the seed for sowing (selection, drying, storage, and pre-sowing seed treatment). Thus they were the masters of the farm management situation without external dependencies and exploitation. The traditional practice had relevance to their skill, their life, their environment, and their mode of production, and was horizontally integrated with their total way of life.

Biogas Plants

Even though most of the difficulties are common to gobar gas and biogas plants, there are also certain peculiar problems related to the latter.

1. In the first place, there is a stigma attached to the gas produced from human excreta. Even the well-informed and educated people prefer to use the gas only for heating water, not for cooking.
2. Usually the human waste is not handled directly by all persons except as compost for fields. Slurry clearance and maintenance, therefore, will have to be done only by members of certain castes.
3. The mix for a biogas plant is calculated on the basis of average human waste per head. These standards do not apply to the rural poor who are, in reality, far below the average adult person with respect to intake and waste disposal. In most of the villages the staple food is not rice and wheat, but maize, ragi or other millets. These are consumed in the form of hot or cold porridges, resulting in excrement which contains more water than solid waste and seldom conforms to scientific calculations adopted now.
4. Also the quantity of intake per head is very low. Most of the families have only two meals a day and at times only one meal per day. Hence the quantity of human waste will also be less. One of the respondents stated rhetorically: "Biogas programme is standing on its head; it is a programme designed to make use of the excreta of the empty stomachs of rural India."
5. There is enough space in the rural area for open-air defecation. In order to change the existing habits of the rural population, the government introduced a "community latrine scheme" which is "an utter failure." The community latrines are not part of the felt needs of the rural people and do not form a part of their social life. There were no organized pre-project attempts to educate the people. The scheme also failed because of other administrative shortcomings. The latrines were urgently built in the villages even without the provision for running water, a fact that confounded the problem and made it worse.
6. The rural people leave their homes in the early hours of the morning and return only late in the evening. They answer to

"nature calls" mostly during working hours while being away from home. Thus the households latrine is an "unwanted" annexure.

Rural Industries.

This has always been the realm of appropriate technology. What is needed now is the improvement of traditional methods and the development of new tools and techniques that increase the productivity, reduce the drudgery of manual labour and enhance the purchasing power of the rural people. During the last decade AT has made an impact on the following industries: spinning and weaving, leather industry including curing, preservation and tanning, processing of cereals and pulses, oil extraction, production of matches and fireworks, soap-making with non-edible oils, handmade paper, coir and other fiber industries, bee-keeping, extraction of essence from medicinal and aromatic plants, bottling of soft drinks, making of toys and handicrafts, preparation and packaging of pickles, jam, chips and snacks, and a host of the other traditional rural industries.

Power Ghani (oil press)

The traditional way of oil extraction has been improved through power-driven machinery. The power ghani is fitted with 2 h.p. motors and pestles made of steel or wood. Critics question whether this innovation is truly an appropriate technology because this has replaced labour and animal power with electricity. Earlier there used to be two bullocks driven by one person while another person collected the oil extraction from the mortar. Now only one person is required for the whole operation. Mechanical ability has, of course, increased extraction of oil by 10 per cent. But the traditional ghani which was a flourishing family trade has now been transformed into a mechanized business retrenching 50 per cent manpower and considerable animal power.

Incidentally, many of the beneficiaries under this scheme stated that the shaft used for driving the pestle breaks down very often. However, one of the beneficiaries claimed that the shaft of the Tata Benz lorry fitted in the power ghani supplied through government agencies could be substituted by that of leyland lorry for prolonged trouble-free service. This has not been communicated to the higher authorities and could not be verified. If the claim is true, then thousands of beneficiaries under this scheme are still struggling with an inferior shaft for lack of follow-up thrust.

Potter's Wheel

The age-old potter's wheel has been now refined and fitted with 1/2 h.p. motor which facilitates uniform motion as well as differential drive. The advantages include greater efficiency, increased production and elimination of drudgery, but mechanization has once again eliminated some jobs. Earlier there used to be two men involved in this job: one to rotate the wheel manually and the other to shape the articles. Though one man is now retrenched, the advantages of mechanization for outweigh the disadvantages and the respondents have been enthusiastic about the innovation.

Generally speaking, AT (Appropriate Technology) in rural industrialization is heading for a new "village industry revolution". In khadi, soap industry, bakery, groundnut cake manufacturing, palmghur industry, leather industry and processing of cereals appropriate technology has already made significant advances, and the beneficiaries enthusiastically support the employment-generating and drudgery-reducing technologies. Indeed, criticisms are voiced in some quarters that khadi and soap industries have "over-expanded" to the point where AT will no longer be efficient unless projected expansion precludes further concentration of productive factors and initiates a deliberate decentralization drive.

Even in rural industries where ATs are well received and are working successfully, the software is grossly inadequate. The system of marketing may be singled out as the weakest link in the chain of AT software. Even where rural industrial products are competitive both in quality and price, they find it difficult to make any significant dent into modern commodities market which is manipulated by big business through affective media blitz and "consumer research". Rural industries have to take a leaf from modern market research, not only to cater to consumer interests but also to generate them wherever appropriate.

A surprising finding in this connection relates to the so-called TRYSEM (Training Rural Youth for Self-employment) programme. Although there is a general agreement on the inadequacy of AT software in India, the curriculum at the Rural Extension Training

Centres under TRYSEM is not significantly different from that of the Urban Industrial Training Institutes where blue-collar workers for modern industries are trained. In fact, their syllabi do not contain any instruction in AT at all. Is it any wonder, then, that the concept of AT is so completely misunderstood by the planners and beneficiaries alike?

Smokeless Chulah (Improved Cookstove)

The traditional cookstove which has remained the same for hundreds of years all over the Third World is a highly inefficient system which often uses less than six per cent of the energy released by the fuel whereas over 90 per cent of the heat goes to waste. The advantages of the new cookstoves are too obvious to be repeated here. There are several types of cookstoves developed in India and for at least three decades they were being propagated in different parts of the country. Although there are significant individual success stories there are probably not even 10 villages where this simple technology has been widely adopted and integrated into the rural social and economic structure. The reasons are many. The cookstoves cost anywhere from Rs. 100 to 300. This initial investment is not a paltry sum for many of the households in the rural area but with a subsidy from the government the technology is within their reach. But why spend any money at all on an invention which is "unnecessary" if not absolutely bothersome? The improved cookstove is not efficient if we do not make use of the double-hearth; most village women say they can manage with one "burner". Men consider the new cookstove a "luxury" because they seldom experience the problem of fuelwood gathering and assume that their "women always manage that". Moreover, to people who live in thatched huts smoke is an insurance against termites, scorpions and even snakes. The respondents also have complained about design and structural deficiencies of the stove and the exhaust pipe.

Agricultural Reforms

The twin purposes of the AT in agriculture are to grow more food and to make rural labour more productive and less arduous. But these objectives are to be met by tying together the two most vital components of agriculture – the human and the environmental – and by ensuring that the yields are sustainable over long periods of time without overtaxing the land's capacity to continue to produce for succeeding generations and without disturbing the ecological balance. With these objectives in view a number of new technologies have been introduced into Indian farming which falls into two general categories: (a) Natural resources improvement and (b) Human knowledge and skills improvement.

Under the natural resources improvement schemes, better irrigation techniques, land reclamation measures, pest control activities, improved tools and techniques of farming, soil conservation measures, drainage, soil upgradation, dry farming techniques, social forestry, farm forestry, mixed farming and a host of other projects were launched. Under the human knowledge and skills improvement schemes, Training and Visit programmes, lab-to-land programmes and other extension education projects were initiated. These projects have involved literally millions of beneficiaries and thousands of extension workers, and the investigator could interview only a few. Although many of the programmes are not packages of new knowledge, they are appropriate development technologies designed to increase production and improve productivity of rural labour. The survey pointed to a few significant pitfalls in the design and implementation of the new projects. For example, the contour bunding schemes which ignored the traditional land-holding patterns and sought to errect new farm boundaries based on technological and topological considerations met with stiff resistance. The scientific norm of a natural farm unit conflicted with the traditional norm of land-owning system. The landholders feared that contour bunding will erase the boundaries and create conflict with regard to land ownership. Even though Punjab was able to progress through the 1960's by means of statutory provisions the contradictions arising out of self-interest, security and public good gave risk to resistance in the South. What is surprising is that the government agencies and extension workers were surprised at the opposition when the incompatibility of the recommended practice and the existing land-owning system was so obvious. Indeed, the resistance was so great that, in the absence of any alternatives, contour bunding was often carried out with individual landholdings rather than natural divisions as the basic units. This is a glaring example of the failure of the extension machinery to take into account the social and cultural factors before attempting to transfer science and technology to rural areas.

Training and visit as well as Lab-to-Land programme have played a crucial role in agricultural development in recent years. The lack of follow-up thrust has emerged as the main lacunae in these and other agricultural extension projects. Under the social forestry scheme millions of tree saplings have been given away absolutely free to almost anyone who requested for them but there have been no significant follow-up efforts to determine how many of these saplings have been nurtured and protected. The communication pattern in the Training and Visit and Lab-to-Land programmes is serving as linkage between the local community and research units. Here again, there is no follow-up action in the programme to see that the knowledge thus disseminated is truly assimilated by the agricultural community. The farmers often do what they have been advised to do during every season for each crop. But this is more like firefighting than organized development. Dissemination of knowledge is a function of the extension agency but assimilation of knowledge is culture itself. If the application of science and technology to rural areas is to become a permanent reality, development planners and extension workers must become fully cognizant of the cultural nexus of rural communities, not simply the economic viability or technical feasibility of any innovation.

Renewable Energy Sources

Windmills and solar cookers are two major AT innovations in this area. Development planners are trying in a big way to propagate windmills that are locally adapted to suit the low-wind conditions generally found in India. These windmills can operate tube-wells as well as lift water from open-dug wells up to a depth of 20 meters driving a device similar to a simple hand pump. Yet the dissemination of windmills in India faces serious problems. In spite of substantial subsidies from the government, the cost of installation and maintenance is too prohibitive for the marginal farmer. Also watering during crucial period of plant life could not be risked, left to erratic winds. Further research is needed to develop light weight windmills suitable for domestic purposes.

The experiments with solar cookers tell a similar story. Although solar cookers have been introduced in some areas with much fanfare, they do not fit into the rhythm and routine of the villagers' life. The villagers cook their meal (often they warm up the leftovers from the previous night) before sunrise and leave for the field. They return only after sundown to prepare the main meal. Moreover, the cooking habits and the nature of preparation of different dishes which are to be mixed and spiced at regular intervals make the solar cookers impractical. At present the cost is also prohibitive and the farmers cannot afford the "fancy gadget" which, in the first place, does not meet their needs.

Conclusions and Recommendations

There is no doubt that AT (Appropriate Technology) has made significant advances in many realms of development in India. However, there are certain fundamental problems associated with software and delivery systems that require immediate attention.

Research in AT hardware is fairly well advanced in India and technological capabilities exist in a great measure. But AT software development is plagued by numerous problems. Take solar cookers and improved cookstoves, for example. Every Indian Institute of Technology, Agricultural University, Polytechnic Institute and Rural Development centre seems to have developed its own version. There is considerable duplication of research and literally hundreds of models clutter the labs. But how many of them are being utilized by rural people for whom they are intended? The plain truth is that even where these technologies are "field tested" there is no effective follow-up. The flow of information is always one-sided – from research labs to rural communities. The practical problems confronted by the beneficiaries are seldom communicated to the research labs in their entirety. Even the field test is often undertaken to determine the technical efficiency of the innovations. It is thus assumed that an inherently superior technology will automatically succeed. Experience has shown that many innovations fail not for want of technical efficiency but because of the lack of social efficacy. The technology must fit into the system of social life and be compatible with cultural traditions. These aspects cannot be "field tested" in a matter of weeks or even months.

A fundamental solution to this problem is beneficiary participation in the design and development of technologies. This will necessitate a reversal of the procedure now being followed. Instead of developing technologies that scientists consider essential and appropriate, we must first ascertain from rural people their needs and requirements. There are, of course, limitations and exceptions to this rule. For instance the requirements of sanitation and public health cannot be left to the whims of the villagers but here again a strategy that begins with the people's participation is far more effective than the one that takes it for granted during the implementation stage.

Any effort to enlist people's participation must, of course, be preceded by an intensive extension education campaign. This means that the successful dissemination of any innovation requires a carefully planned pre-project phase which should include at least three strategic steps. The first step involves the classical community development principle of identifying the felt needs of the rural people and planning solutions for them. The second is to educate the rural people to identify the real needs. The outdoor privy, improved cookstove or better food processing techniques may not be easily identified as the real needs and it is left to the extension agency to help the people define these needs and evolve solutions in collaboration with them. The third strategy should draw upon modern marketing techniques and consumer psychology. Modern marketing research has shown that an appropriate psycho-social condition can be created where the consumers will be eagerly looking for the launching of a certain commodity in the market. While this may be particularly true of the urban middle class, research has indicated that rural people are not far behind in adopting innovations which are valuable to them. During the pre-project phase the innovations should be introduced into the psycho-social dimensions of the rural society through education, creating awareness, implanting the desired conceptual framework and effective feed-back. Education should play a prominent role throughout the pre-project phase which in some cases may be prolonged especially if the innovations involve the alteration of any social or economic structures or cultural patterns. As we saw earlier, soil conservation measures and contour-bunding schemes faced stiff resistance because of farmer's fears and suspicions about the government's intentions and there was no pre-project phase to educate and win them over.

An analysis of the responses to questions pertaining to software consistently pointed to a major lacunae in the feedback system. In field trials and follow-up the scientists and technicians have been primarily, at times almost exclusively, concerned with the technical efficiency of the innovations and to a certain extent their economic viability. Questions pertaining to social acceptability and cultural adaptability were either ignored or given only cursory attention. With the recent technological advances technical efficiency of the hardware is a relatively simple matter. What is really needed is an overhaul of the software and an enlightened sensitivity to the social and cultural fabric of the rural communities.

Another pertinent question concerns the role of social scientists in the development, promotion and application of appropriate technology. In almost all of the Rural Development Centres and Technological Institutes AT has been almost exclusively in the hands of technologists and engineers. Social scientists, except for a handful of rural economists, seem almost totally oblivious of the AT movement. Very few sociologists, anthropologists, or psychologists seem to have any interest in this crucial development strategy. It should, therefore, surprise no one that the sociocultural and psychological factors that facilitate or impede the diffusion of AT remain largely unexplored. It is high time that social scientists team up with their counterparts in science and technology in the research and development as well as application of AT to rural areas.

The present enquiry also confirms the oft-repeated assertion that the government bureaucracy is neither professionally trained nor psychologically oriented to handle AT as a science and technology movement. The government functionaries are still target conscious and often primarily concerned with the number of projects to be completed during the plan period. Some of them also have a tendency to treat the development projects as "hand-outs" or "welfare schemes" which must somehow be delivered. Very few of the government functionaries are trained in the philosophy and practice of AT. This explains why the confusion about the concept of AT persists even among the extension workers and why many of the AT projects are construed by the general public as welfare programmes or mere political trumpetry. The conception of AT as a science and technology movement adapted to the needs and conditions of rural societies is almost blatantly absent. AT must become an integral part of the curriculum at all levels, both in formal and informal education, and especially of the training programmes and orientation for rural development workers and community leaders.

Hardware specialists must collaborate with one another, share technical information and data, avoid unnecessary duplication and supplement the resources. Software specialists must develop an appropriate functional language that is less technical and more practical. Both hardware and software specialists need the experience and expertise of trained social scientists who can provide insights into the social and cultural framework of target groups and help determine the appropriateness of any AT products or methods. It is high time that social scientists look beyond abstract social systems and predictable phenomena, and get involved in the business of AT-the grassroots science and technology movement which is so inextricably intertwined with the process of nation-building in developing countries.

References

Anil Agarwal. Solar Cells: Plugging into the Sun. Center for Science and Environment monograph (CSE), New Delhi, 1983.

———, Firewood: Fuel of the Rich. CSE monography, 1983.

———, China's Habitat I: Use Mud, Build for All. CSE, 1983.

———, China's Habitat II: Helping People to Help Themselves CSE report, 1984.

———, China's Habitat III: Urban Housing: A Total Disaster, CSE report, 1984.

———, Mud, mud: The potential of earth based materials for Third World Housing, Earthsan, 1981.

Anil Agarwal and Bhubanesh Bhatt. Firewood in Cities I: The dimension of the rural-urban firewood trade. CSE report, 1983.

———, Firewood in Cities II: The impact of the rural-urban firewood trade on the people and forest of Madhya Pradesh. CSE report, 1983.

Anil Agarwal and Priya Deshingkar. Headloaders: Hunger for Firewood--1. CSE report, 1982.

———, Headloaders: Burdens of Energy--II. CSE report, 1981.

AFPRO. Decentralized Energy Planning. New Delhi: Action for Firewood--1. CSE report, 1982.

———, Biogas Panel meeting: Proceedings, 1981.

———, Workship on Biogas Technology Extension, 1983.

Peter Askins. Intermediate Technology: An Informal Survey. Washington D.C.: 1976.

Asian Development Bank. Appropriate Technology and its Application. Manila, 1977.

ASTRA: Proceedings of the ASTRA Seminar. Banglore: Indian Institute of Science, 1983.

D.R. Bhumbla. Canal Irrigation: Impact on Production and Environment. CSE reports (103 & 104).

Marilyn Carr. Appropriate Technology and Rural Industrialization. ITDG, 1982.

Central Projects Staff. Appropriate Technology and World Bank Assistance to Poor. Washington D.C.: World Bank, 1978.

Ujjayant Chakravorty. Maize: Waiting for the Next Green Revolution. CES Report, 1982.

Kamla Chowdhry. Managing Forests for Development: I & II. CSE report, 1982 & 83.

———, A Green Revolution that Begins with Schools. CSE report, 1984.

R.J. Congdon (ed). Introduction to Appropriate Technology. Emmaus, PA: Rodale Press, 1977.

Consortium on Rural Technology. Cookstoves--Firewood Burning, Designing & Transfer Workshop Proceedings, New Delhi, 1983.

———, Rural Sanitation: Technology Options, 1981.

———, Solar Cooker and Ovens: Technology Options, 1981.

Ken Darrow and Rick Pam. Appropriate Technology Sourcebook, Vol. I Stanford: Volunteers in Asia, 1976.

Ken Darrow, Kent Keller and Rick Pam. Appropriate Technology Sourcebook Vol. II.

Stanford: Volunteers in Asia, 1981.

Department of Science and Techology. Annual Report 1983-84. New Delhi: Government of India.

———, Science and Technology for Women. New Delhi, 1982.

Priya Deshingkar. Solar Cookers: A Ray of Hope? CSE monograph.

Anil K. Dhussa. Designing the Biogas Distribution System. AFPRO monograph.

P.D. Dunn. Appropriate Technology: Technology with a Human Face. London: Macmillan Press, 1978.

Donald D. Evans and Laurie Nogg Adler. Appropriate Technology for Development. Boulder: Weestview Press, 1979.

Peter Fraenkel. The Power Guide-- A catalog of small scale power equipment. ITPL, 1979.

Madhav Gadgil and Ravi Sharma. Ecology is for the People. CSE report, 1981.

M.K. Garg. Some developments of appropriate technology for improving physical amenities in rural homes. ATDA, 1978.

———, The development of an appropriate technology for decentralized pottery industry in rural India. ATDA 1980.

———, Pilot project of decentralized spinning technology for handloom weavers. ATDA 1976.

———, Feasibility report on mini-sugar technology. ATDA 1979.

———, Mini cement technology. ATDA 1980.

GATE. Low Cost/Self help Housing. German Appropriate Technology Exchange, 1980.

———, The Utilization of Renewable Energy Resources - A program of the Federal Republic of Germany for the Third World, 1982.

———, Low Cost Transportation, 1979.

Rajiv Gupta. A Sunny Lunch? CSE monograph

———, Finding the Niche for Wind Energy. CSE monograph.

———, Putting a Tap to the Wind. CSE monography.

———, Fire Without Smoke. CSE report.

———, Animal Power: As important as Electricity. CSE report.

———, Return of the Bike. CSE.

———, From Waste to Resource: Water Hyacinth. CSE report.

———, Fathh Singh Ka Purwa: A Community Venture. CSE report (on community biogas plant).

———, When the Hills Came Tubling Down. CSE report.

Malcolm Harper and Tan T. Şoon. Small Enterprises in Developing Countries. ITPL, 1979.

M.M. Hoda. Directory of machines, tools, plants, equipment, processes and industries. Lucknow: Appropriate Technology Development Association (ATDA), 1979.

———, Rice Milling, ATDA 1981.

———, A proposal for development of Appropriate Technology in India, 1980.

———, Appropriate Technology and Research Projects ATDA, 1974.

———, Solar Cookers. ATDA, 1979.

M.M. Hoda and S.C. Bhatnagar. Directory of Agricultural Tools and Impliments. ATDA 1984.

International Bank for Reconstruction and Development. Village Water Supply and Sanitation in Less Developed Countries. Washington D.C. 1974.

S. Jackson, Economically Appropriate Technologies for Developing Countries.

Washington D.C. Overseas Development Council, 1972.

Nicholas Jequier. Appropriate Technology: Problems and Promises Stanford: Volunteers in Asia, 1976.

Balraj Kumar. The Dung-carrying Donkeys of Agra. CSE report.

John Madeley. Rediscovering the Forgotten Crops. CSE report.

Ken Marshall. Package Deas: A Study of technology development and transfer. London: Intermediate Technology Publications Limited, 1983.

George McRobie and Marilyn Carr. Mass Production or Production by the Masses. London: ITDG. 1982.

R.C. Mangal and Narendra Varma. Achievements in the Field of Rural Buildings and Environment. Roorkee: Central Building Research Institute, 1979.

D.W.J. Miles. Appropriate Technology for Rural Development: The ITDG Experience. London: Intermediate Technology Development Group (ITDG), 1982.

Ward Morehouse (ed). Science, Technology and the Social Order. New Brunswick: Transaction Books, 1979.

A.S. Myles and Ramond Myles. Donflow Design Concept of Biogas Generation from Water Hyacinth. Action for Food Production (AFPRO), New Delhi, 1983.

Raymond Myles. Pilot Demonstration Project on Low-Cost Night-soil Biogas Plants for Community and Family Use. AFPRO monograph, 1984.

————, Training of Village Masons on a Low Cost Biogas Technology. AFPRO monograph 1984.

Sevanti Ninan. Chinese Biogas Program Sputters. CSE report.

Arnold Pacey. Rural Sanitation: Planning and Appraisal. ITPL 1980.

Mohan Parikh. Agricultural Tools Research Centre. 1980.

Tram Praudyogiki (Center for Rural Development and Appropriate Technology) Journal. New Delhi: Indian Institute of Technology (especially 1982 & 1983 issues)

Preliminary Report from a Workshop on Appropriate Technology, Morrilton, Arkansas, Winrock International Conference Center, 1978.

Jairam Ramesh and Charles Weiss, Jr. Mobilizing Technology for World Development. New York, Preger, 1979.

Shrikanta Rao. Village oil industry. ATDA. 1982.

Rathindra Nath Roy. Integrated Systems; Integration, application and management of energy actvities at the community level: technology transfer and communication in R.C. May et.al. (eds).

Appropriate technology for alternative energy resources in fisheries. ICLARM Conference proceedings, Manila. 1982.

W.H. Schacht. Appropriate Technology: Technology for Growth in the Developing Nations. Washington D.C.: Congressional Research Service, 1978.

E.F. Schumacher. Small is Beautiful. New York: Harper and Row, 1973.

Ravi Sharma. Traditional Bullock Carts: Has the Decline Finally Begun. CSE report.

————, Blue-Green Algae: Heralding in a Rice Revolution. CSE report.

————, Greening the Countryside. CSE report.

————, Preserving the Flora. CSE report.

————, Final Warning to Asian Countries, CSE report.

————, Rethinking Big Dams: I & II. CSE reports

Shri AMM Murugappa Chettiar Research Center, Madras. Technical Notes No. 9, 1983.

Hans Singer. Technologies for Basic Needs. Geneva: International Labor Office, 1977.

J.B. Singh. Biogas Program: Involvement of Voluntary Organizations. AFPRO monograph.

SKAT. Manual for Rural Water Supply. Swiss Appropriate Technology organization, 1980.

Ram Swarup. Gandhian Economics: A Supporting Technology. ATDA, 1977.

UNESCO. Technology for Rural Development--Modalities for action. Report of Experts Meeting, Brussels, 1980.

U.S. Agency for International Development. Proporal for a Program in Appropriate Technology. Washington D.C.: U.S. Government Printing Office, 1977.

VITA. Training Village Entrepreneurs--Guidelines for Development Workers. Mount Rainier: Volunteers for International Technical Assistance, 1980.

———, Village Technology Handbook., 1981.

———, Environmentally Sound small-scale agricultural projects, 1979.

John Vogler. Work from Waste-Recycling waste to create employment. ITPL, 1981.

5

DEVELOPMENT AND SOCIAL AWARENESS
Perils of Pesticide Use In Modern Agriculture

C.R. PRASAD RAO & SUSAN VERGHESE

Introduction

The overwhelming preoccupation of development literature is mostly with questions of "growth," "equity," "strategies," "hurdles," "infrastructures," "institutions," and "grass-roots planning and participation." While the social, economic, and political consequence of capitalist and socialist models of development have generally received the most attention, a whole range of other outcomes of the developmental process are rarely discussed. Neglected for a long time, for instance, were the issues of long-term ecological and environmental concerns – conservation of natural resources, preservation of ecosystem, environmental pollution, public health hazards, waste disposal, and industrial disasters. Also bypassed were the immediate issues of ethical, legal, organizational, and monitoring problems of development moving on its set course. Viewed in the broadest terms, the culture lag theory in sociology relates to the disjoints and disturbances that any significant technological change generates in the institutional, organizational, and behavioural realms of society. In the absence, or retarded development, of complementary adjustive changes in values, organizational mechanisms, and behaviour patterns, technological change often proves counter-developmental. Given their technological bias, the technocratic sponsors of development reverse this position by holding that disruptions in prevailing institutional arrangements are prerequisites of sustained and effective technological modernization. Bhalla and James maintain,

> "The introduction and absorption of new technology will inevitably shake up prevailing institutional arrangements, and if the disruptions are significant, alterations in attitudes, cultural values and norms of behaviour will tend to become more flexible and receptive to further technological progress. This thesis, stressed by the Thorstein Veblen/Clarence Ayres axis of institutional thought, is so familiar...." (1986:456).

The inherent danger of this thesis is that once technology creates for itself a receptive climate, the ground is prepared for axiomatic acceptance of both desirable and deleterious technological innovations. The technological basis and bias of the modern, complex development is no doubt currently being questioned, but only in non-fundamental instrumental-rational terms of its relevance, cost-effectiveness, and appropriateness vis-a-vis the development objectives of production, productivity, income enhancement, employment generation, distributive justice, etc. Ironically, the relevant critical issues concerning the technological thrust of modern development are once more addressed and phrased in technological terms (e.g. alternative or intermediate technology or super technologies). In other words, solutions for technology-created problems are sought within the confines of technology, thereby underscoring and legitimating the technological bias of modern development.[1] This may be a result of the modernist obsession with the presently evolved technology, or a subtle outcome of the technocratic vested interests, or a byproduct of the march of the history away from traditional systems of life and production–a race against time! The potentials of modern science and technology to rediscover and redeploy the time-tested traditional systems are not harnessed. For example, very few are the persons and groups who will not be disinclined to perceive a direct relationship between the modern technologies and the eruption of the social evils of hijacking, terrorism, spurious drugs, and pernicious pesticides. The technocratic argument easily dismisses these evils as ethical, legal, or administrative problems to be tackled on a different front, rather than admitting them as inevitable, logical and inexorable concomitants of technology development. Naive and facile is the argument that holds that technology itself is neutral, since a morally neutral technology can only exist in a vacuum. By its nexus with society, technology easily transmutes itself into a destructive one by human greed, design imperfections, impulsive actions, judgment errors, technical mishaps, terrorism, sabotage, deliberate enemy action, industrial disaster, and natural calamity. As Srinivas Iyengar (1984:17) writes:

> "... one or more or all of them start Operation Doomsday and encompass the inglorious end of the human adventure on earth. Science and technology being amoral, scientists and technologists being under the control of shortsighted but ambitious politicians and militarists and industrialists, it is too much to hope that *the technological race will halt altogether or its dangerous deployment will be eliminated entirely.*"

By themselves science and technology cannot provide from within controls and correctives against their abuses by the political, military, industrial, and bureaucratic elite who "sport an omni-competence that awes the common people." The human element and equation saps away the thin veneer of neutrality, and given its all manner of potential deployment, technology becomes the progenitor of perverse and perilous policies. This technological determinism implies that certain policy options emerge as the invariable results of the available kinds of technology. This is the spirit which informs the tenor of this paper on development and social awareness, taking the use of pesticides in modern farming as an illustrative case. The paper pulls together evidences from various reports and sources in order to concretely focus sensitivities upon some issues of the modern technology-based development. A concern which underlies the whole paper is to persuasively demonstrate that several of the culture lags and other challenges of technological innovations are not easily solvable and may even spell disaster too soon to allow amends and corrective measures. If there are real limits to progress it is better that the mankind recognizes the "whats," "whys," and "wheres " of those limits. Extreme technologism is as reprehensible as extreme obscurantism. Proper cultural policies must guide and regulate the nation's development policy (Joshi, 1983; and Kannan and Ramanujachari, 1986).

The Pesticide Problem

Pesticide use, a critical modern agricultural input, has become an established global practice. Together with synthetic fertilizers, it is increasingly chemicalizing agricultural production and agroecosystems. Its place in the yield-augmenting package of agricultural practices is only as a yield-saver, for it does not contribute directly to crop yield enhancement. It contributes merely by cutting crop losses caused by insect pests and diseases thereby ensuring a realisable high yield under the farmer's particular agro-climatic conditions. The loss-elimination by pesticides accounts for up to 30 percent of the total produce in normal conditions. Loss elimination is a boon to the farmer using high-cost agricultural inputs – chemical fertilizers, artificial irrigation, mechanized cultivation, and/or heavy hired labour. From pre-sowing stage to harvest, and post-harvest storage of the farm produce, the farmer has to confront the problem of pests and diseases. The problem is particularly acute when highly vulnerable high-yielding varieties of crops are grown. Irrigation itself brings its own pest problems by facilitating the breeding of insects, the multiplication of weeds as alternative hosts, and the multiple cropping. Cropping schedules or the sowing and harvesting of a crop at different times by farmers of contiguous fields makes the area liable to the risk of pests and diseases which simply migrate from one field to another. Similarly, late sowing of crops due to non-availability of seeds or irrigation water or rainfall invites the pest problem. Skipping of fallowing and rotation of crops also has the same effect.

The losses from pest attacks may sometimes be total (e.g. growing of monocultures and attacks by grass-hopper swarms). A virulent attack last year of white fly on cotton grown in two districts of Andhra Pradesh has wiped out or rendered unsaleable almost the whole crop.[2] The loss incurred was so heavy that in the current year the area grown under cotton was reduced by about 45 to 50 percent of the last year's acerage.[3]

The race for higher yields of crops in modern farming has caught the farmers simultaneously in a race for ever-increasing application of pesticides. But these two races have resulted in a third race namely, the emergence of resistant strains among old pests and newer insect pests. Pesticide sprayings often create vermin troubles where none existed before (Dover, 1985: 52-61).

The use of pesticides in India has increased from 2,000 tonnes a year in 1950 to 72,000 tonnes a year in 1983.[4] In the year 1985, the use of pesticides in India's agriculture was of the order of 80,050 metric tonnes and by the end of the Seventh Five Year Plan it is expected to go up to 93,700 metric tonnes.[5]

During the period 1972–83, global sales of pesticides rose from $ 8.1 billion to $ 12.8 billion, with the most rapid growth in the Third

World (Crump, 1986: 1). The USA accounts for 50 percent of the world's aggregate use of pesticides; 6 billion pounds, worth $ 3 billion were used in 1975.

Countrywise, Japan uses 10,000 gms., Europe 2,000 gms., USA 1,500 gms., and India 400 gms. per hectare (Sharma, 1985: 3).

Non-availability of data by state, crop, and farmer class places us at a disadvantage in presenting a picture of the variations and methods of use of pesticides in India's agriculture.[6] It has been mentioned early that the inter-linked character of the package of practices in modern farming methods and the cost of production influences the extent of application of pesticides. In general, the higher the cost of other inputs, the greater will be the use of pesticides by a farmer. Nevertheless, pesticide consumption in India is not as high as in some advanced countries.

In general, three main strategies of pest control are available to farmers: (1) control by chemicals; (2) control by biological methods, i.e. predator insects; and (3) integrated pest management which includes chemical, biological, mechanical and cultural means of pest control in which the use of chemicals will be the lowest. Historically, control by chemicals is the youngest of the three strategies, and it has overshadowed the other two strategies in its overall appeal to farmers as well as its impact on yields and agroecosystems. (Krishna Murty, 1983: 15-16). Estimates of the available range of basic chemicals and pesticide formulations differ widely. According to one estimate, there are 1,500 basic pesticide chemicals and 50,000 formulations (Sharma, 1985: 3).

Harmful Effects

Several evils are associated with the widespread use of synthetic pesticides in agriculture. In the long run, the multifarious harmful effects of chemical pesticides outweigh their benefits. Social awareness of these harmful effects has begun to develop among agricultural scientists, pesticide manufacturing industries, and policy makers. But farmers and consumers are by and large unconcerned, if not entirely unaware. The prospect of vigorous and sustained movement against chemical pesticides is remote and dim at present. In so far as agricultural scientists are concerned, this awareness is abundantly reflected in the latest shift of emphasis from prophylactic use of chemical pesticides to their need-based application and integrated pest management strategies. Farmers are now being asked to use chemical pesticides *sparingly, strictly* according to recommendations, and only when the pest populations have reached the *critical size*. Given the present demand pattern for chemical pesticides and their appeal to farmers, it is difficult in practice to ensure that the farming population's pesticide use behaviour conforms to the above specifications of need-based, strict, and sparing use of pesticides. The practical implications of these recommendations in the socio-economic context of the Indian farming community must be examined clearly and systematically by actual field empirical studies of farmers' pest control behaviour from pre-sowing stage to marketing.

The harmful effects of chemical pesticides can be considered from different angles and in different domains. (1) They can be examined in relation to their effects on the physiology of crop plants and the toxicities of pesticide residues in crop products, particularly foodstuffs and livestock fodders. (2) They can be considered for their effects on the agroecosystem, including the soil, the non-target insect species, and predators and soil-borne organisms, and in terms of the properties of persistence, degradation, and resultant interaction of chemical pesticides with other chemicals in the agroecosystem. (3) The pesticide evils may also be discussed with reference to genetic mutation and resistance within target insects and the resultant emergence of new resistant strains. (4) Finally, the pesticide problem is relevant from the point of view of the health of the agricultural workers engaged in the spraying of chemical pesticides, the consumer's health, and the damage to workers in the pesticide industries.

Resistance to Pesticides

Although resistance to pesticides in insects was noticed as early as 1911, their large-scale application in recent years has accelerated the pace of development of insect resistance. Owing to years of exposure, many pests have become immune to pesticides. In the field, the spraying of chemical pesticide affects both the target and non-target insects, all of them developing resistance which may be specific or general. The biological mechanisms of resistance among insects to pesticides are fairly well-known and are being explored further. For instance, certain enzyme secretions by the insects may decompose the pesticide and destroy its killer power. The insect may also develop resistance by altering the rate of penetration of the insecticide or the site of the pesticidal action. Krishna Murty (1983:15) reports:

> "It is well known that the use of given pesticide on pest population can result in the selection of individuals with tolerance to concentrations of the chemical much larger than needed to kill the majority of the pest population. These strains breed and produce resistant populations. Mites, ticks, fungi and rodents have all shown resistance to one pesticide or the other. Given the necessary selection pressure, it is feared that soon all pests may turn resistant to every available pesticide. The consequences of such an exigency on world health and world food production could be disastrous."

Weeds have become resistant to weedicides. Michael Dover (1985) reports that from 1970 to 1980, the number of resistant species of mites, ticks and other insects jumped from 224 to 428 – an upsurge that threatens U.S. cotton and potato farming. The Food and Agriculture Organization's Panel of Experts on Resistant to Pests lists 223 agricultural pests which have become resistant to nine of the major group of pesticides. These pests are of importance to global economy (Krishna Murty, 1983: 16). Sharma (1985: 4) holds:

> "...In USA the pesticide inputs increased 12 times in the last 30 years but the pre-harvest crop losses doubled, a pointer to the pest resistance, new pests and crop behaviour."

Pesticide Toxicity Effects in the Food chain

Chemical pesticides are known to produce stress effects on biological systems, some of them toxic to the genetic material. Man is exposed to these directly through contact with the chemicals or indirectly through the food chain. Toxicity to the genetic material may occur at the genic, chromosomal, and/or genomic levels, leading to carcinogenesis (cancer), teratogenesis (malformations and monstrosities) and hereditarily transmitted diseases (Sharma, 1985:1). The toxic effects of only 17 percent of the large number of pesticides are at present investigated. Out of those investigated, 70 percent were found to be genotoxic in one or the other system. In other words, 90 out of 126 tested pesticides were genotoxic. Nearly, 50 percent of all the studies were from plant systems. Thus there is a gaping void in our knowledge of pesticide toxicities. A lot more pesticides are in use whose toxicities are yet undetermined down the food chain (Sharma, 1985:4). In plants like wheat, barley, and sorghum, cytogenetic abnormalities, loss of fertility and chlorophyll deficiencies, and emergence of normal weeds were traced to genotoxic effects of pesticide stresses. A higher rate of mutations (genetic changes) among the organisms lower in the evolutionary ladder, along with a trend of decreasing crop resistance, is expected as a result of pesticide stresses (Sharma, 1985: 8).

> "It is known that both the resistance to pesticides (in) the pests and susceptibility to newer pests (in) the crops are gene controlled and therefore inherited." (Sharma, 1985: 8)

As regards health hazards of pesticide residues in human beings, the Natural Resources Defence Council, U.S.A. conducted a survey during 1983-84 into pesticide residues in food. It was revealed that 44 percent of fresh fruits and vegetables sampled contained residues of 19 different pesticides, several of which are known or suspected carcinogens. More deplorable is the observation that analytical methods used by the Government's Food and Drug Administration to test for residues are incapable of detecting two-thirds of commonly used pesticides (Crump, 1986: 1). In India, according to the Indian Agricultural Research Institute's scientists, the state food testing laboratories attached to the Departments of Prevention of Food Adulteration do not have the sophisticated instruments required to determine the pesticide residue levels in foodstuffs. Many of these instruments are to be imported, the analytic procedures involved in testing are lengthy and laborious, and the instruments used must be highly sensitive (Girimaji, 1985: 11). The overall situation relating to pesticide poisoning of foodstuffs presents the following features: (1) we have no proper means of detecting pesticide toxicities; (2) we have no knowledge of the toxicities of a very large number of chemical pesticides currently in use; (3) from what little work was done, the evidence points to pesticide poisoning in the entire food chain, even in human milk and placenta.

Andy Crump (1986: 1) reports about an unpublished WHO study by Dr. R.S. Levine of the University of Miami's Department of Epdemiology and Public Health. The study was based on information collected from 18 countries for 1983, including Australia, Brazil, Canada, France, West Germany, Hong Kong, Hungary, Yugoslavia, and the USA. From a total of 750,000 reported cases of food poisoning in 1982, 13,800 fatalities were recorded. In 1983, there were an estimated 40,000 deaths from between 1.5 million and 2.0 million reported cases of poisoning. Most of the poisoning cases are attributed to agricultural chemicals, and 50 percent of all cases and 75 percent of all fatalities occur in developing countries.

A survey by the Central Food Laboratory, Calcutta, published in 1980 reports that out of 400 food samples tested, 95 contained pesticide residues, the toxicity levels in 35 of them exceeding the prescribed limits. The National Science Academy Study of 1965 detected DDT residues in eggs, milk, oils, and oilseeds sold in Delhi markets. Twenty-five egg samples collected from Bombay markets showed high incidence of pesticide contamination. As many as 655 samples from a total of 1316 samples of cereals, pulses, oilseeds, grapes, milk, egg, fish, and meat collected from Hyderabad markets were found to contain DDT residues far exceeding the safe limits (Girimaji, 1985: 11).

Legal measures such as sections 65 and 44 (A) of the Prevention of Food Adulteration Act have remained largely ineffective for lack of proper enforcement and lack of technical guidance on the methods of testing foodstuffs for pesticide residues. Further, there is no machinery for monitoring and regulating pesticide usage behaviour of farmers (Krishna Murty, 1983: 1). Uninformed, indiscriminate, and injudicious use of pesticides by farmers entails the risk of pesticide poisoning in foodstuffs brought to markets. In their anxiety to cut losses and save crops, or due to ignorance and illiteracy, farmers violate the guidelines of pesticide application and resort to more frequent applications or higher concentrations than recommended. Sufficient time intervals may not be allowed between pesticide sprayings and harvesting and transporting to markets. In an agrarian system dominated by millions of poor and semi-literate agriculturists backed by feeble and indifferent extension education institutions, and with the virtual absence of monitoring of the farmer's pesticide application behaviour, neither the use of chemical pesticides nor integrated pest management can be properly enforced. This is in addition to the problem of cheating and spurious pesticides manufactured by industries.

> "The prospect of developing safer pesticides which will retain efficacy against pests, but will not disrupt ecological equilibrium and will be easily degradable, does not, however, appear to be bright, if only one takes a realistic view of the frustrations in the area of research in chemotherapy." (Krishna Murty, 1983: 16).

While it may be assumed that safe pesticides can be developed, it cannot be assumed that the safe ways of using them would be established among Indian farmers. A chemical that is fool-proof from all angles is unlikely to exist or be developed. Further, the Third World countries cannot afford to meet the costs and manpower requirements to establish the vast administrative machinery and extension network required to train, monitor, test, and oversee the pathlines of pesticides.

Effects on Agricultural Workers and Ecosystem

By an emergency order, the U.S. Environmental Protection Agency banned and removed from markets a common herbicide Dinoseb and the insecticide Ethylene Bromide. Dinoseb, in common use since the late forties, was found to cause birth defects and infertility in doses of 10 milligrams/kilogram of body weight. Farmers and farm workers who spray Dinoseb are exposed to similar or higher levels of the chemical. Workers who operate sprayers on the ground can receive a daily dose as much as 12 mg/kg and aerial spraying can result in a toxic exposure of up to 39 mg/kg. The chemical affects the fundamental workings of the living cell, and thereby is probably involved in causing cancer and immune system impairment.[7]

A study of the wheat-growing districts of U.P. revealed that one-fifth of the farm labour in the region had weak eyesight due to muscular degeneration. Those engaged in pesticide spraying in malaria eradication programmes showed damage to the nervous system and presence of cancer. A report from the Indian Institute of Occupational Health stated that out of 160 workers engaged in pesticide production, 73 percent showed toxic manifestations, and 35 percent suffered from cardio-vascular and gastronomical complications.[8] In our country neither is there an adequate exposure to such grim health hazards nor is there a public and official concern over the issue. For instance, in our country, 72 percent of the commonly used insecticides consist of BHC, DDT, Aldrine, Chlordane, Heptachlor, Lindane, and Ethylene Dibromide, which have been banned or restricted in other countries on account of their toxicity.[9] BHC, Sodium Syanide and Dieldrin, banned in the Western countries, have been approved for use in India on specious grounds.[10]

The heavy toll of the living systems which excessive use of pesticides takes, leading to acute ecological degradation, is vividly reported by Ramadass' case study of Kerala's rice bowl, Kuttanad

(1986: I-III). Human meddling in Kuttanad seems to have drawn man's ingenuity into an unending struggle with Nature. How Nature avenges man's rapacious efforts is dramatically illustrated by the happenings in Kuttanad's ecology. A summing up of the Kuttanad story here may not effectively portray the events and their interrelatedness in the ecosystem. Nevertheless, it presents a fairly grim picture. Here the fields were originally part of the lake and were reclaimed gradually for rice cultivation by formation of dykes and mud bunds. This fertile rice belt of Kerala is now riddled with serious ecological problems, one main cause being the excessive and indiscriminate use of pesticides. The high-yielding varieties of rice used here have brought along with them new pests. The lack of cropping discipline has also contributed to the proliferation of pests. To counter the heavy pest infestation, pesticides such as Methyl Parathion, Carbofuran, and Monocrotophos, all classified as highly hazardous by WHO, are excessively used. The pesticide sprays cause direct problems such as allergies, asthma, eosinophilia, and skin rashes. It has also resulted in the gradual dwindling of fish and natural predators of crop pests such as frogs and field snakes. The newly constructed barrage which prevents saline water intrusion has also complicated the effects on the ecosystem. A pernicious weed that does not thrive in salt water has made its entry, clogging up waterways. Lack of two-way interchange of fresh water system with sea has caused water to stagnate for most of the year and accumulate organic wastes from human habitation and chemicals and pesticides from agriculture. Mosquitoes breed and multiply in the stagnant waters causing malaria and filariasis. High incidence of gastroenteritis, dysentery, cholera and poliomyelitis is found. None of the official agencies – the Animal Husbandry Department, the Agricultural Department,or the Public Health Department – has any plans to investigate the pesticide residue toxicities in the food chain. The report expresses pessimism that technological fixes to technology-created problems are not real and permanent solutions but will only bring in their wake newer and unknown problems. "Maybe the solution is to call off the battle against nature and live in peace with it."

Conclusion

In the natural ecosystems, various orders of living organisms and plant species exist in naturally induced and environmentally selected systems of mutual and multiple interactions. These interactions, established during a gradual and long process of evolution, are basic to the ecosystem in the sense that they incorporate subtle checks, tender balances, and unknown parasitic, predatory, and symbiotic relationships that collectively sustain the ecosystem as a whole. They constitute the roots of productivity and productive potential of nature. Therefore conservation rather than greedy consumerism and commercialism, preservation rather than perilous exploitation, and harmony rather than harsh disruption of the ecosystems in their natural and pure states, should be the guiding principles of man-nature relationships.

The harmony, peaceful coexistence, and cooperation between man and nature is to be cognized as one dimension of the man-technology-nature relationship. It was mentioned in the beginning that issues concerning technology which mediates man-environment relationships should be addressed in fundamental terms rather than in rational-technologistic terms. A second dimension of the fundamentalist critique of man's involvements with technology relates to the nexus between science and technology policies and ethico-cultural policies of society. If technology causes drastic ruptures in social institutions, should social policy leave technology as an ungoverned, autonomous master of social change? Should technology change be allowed by society a free and unregulated course? The main burden and argument of this paper so far has been to drive home the need for a master of the master of social change. Technology as a potent social force requires another social force to sternly discipline its own course of development. Science and technology have put such a tremendous power at the disposal of society that their course and evolution cannot be left alone to scientists, politicians, militarists, and industrialists. Science and technology policies, no matter how critical and relevant they are to the defence and development needs of society, must not be formulated independently of ethico-cultural values. Rationalism in policy-making needs to be balanced by ethico-cultural fundamentalism. In fact, in the light of the text of this paper, fundamentalism seems to be more rational than rationalism itself. If rationalism is a modern need and value, fundamentalism becomes a supra-national and supra-modern need and value. To dissociate technological planning and policies from cultural planning and policies is to court a disaster of unimaginable dimensions. While in the past it was religion and theology which doubted the adequacy of science, dubbing it as "a false and pretentions Messiah, or at the very least a highly ambiguous element in modern civilization," the current doubts are secular, and humanistic, writes Mills (1959: 15-16).

Science and technology even in their latest incarnations and outfits are experienced as doubtful solutions for the deeper and more fundamental afflictions of mankind by intellectuals, historians, humanists, and culture specialists.

> "They have raised more problems – both intellectual and moral – than they have solved, and the problems they have raised lie almost entirely in the area of social not physical affairs" (Mills, 1959: 15).

Now, science, the chief instrument of the conquest of nature, "the overcoming of scarcity, is felt to be footloose, aimless, and in need of reappraisal", and Mills correctly diagnoses that the technological ethos and imagination are more frightening and ambiguous than hopeful and progressive. "It is the human meaning and the social role of science, its military and commercial issue, its political significance that are undergoing confused reappraisal" (p. 16). Science provides often confused fragments of the realities, lacking in a holistic perspective.

The holistic perspective of science and technology in the present context suggests an integration of the tool-maker role of man which underlies science, technology, and industrialism with his value-maker role (Kannan and Ramanujachari, 1986: 17). The truth values of scientific discoveries and the utilitarian benefits of technological innovations cannot be judged apart from the cultural values of beauty and goodness. All knowledge – rational and scientific or intuitive and revelational – would satisfy the triple criterion of truth, beauty and goodness in the holistic perspective. This perspective, according to Kannan and Ramanujachari, serves as the cultural force necessary to usher in peace and harmony alongside with progress and development. Joshi also perceives the need for an integral cultural framework of science and technology (Joshi: 1983).

Notes

1. For an illustration of the technological bias see, Bhalla and James: (1986: 453-462). The authors visualise intermediate technology and technological blending as alternatives of technological upgradation. They observe, "Regardless of the avenues pursued-intermediate technology, technological blending or spreading pockets of complete modernisation — it is increasingly obvious that some form of technological enrichment is one pre-requisite to improving income and productivity in traditional sectors of the Third World" (p. 455).
2. *The Hindu* April 30, 1986, p. 12.
3. *The Hindu*, January 6, 1987, p. 3.
4. *Science Today*, September, 1986, p. 44.

5. *The Indian Express,* April 25, 1985.
6. Adoption studies conducted by the All India Coordinated Research Project for Dryland Agriculture in 16 centres over four years 1976—80 reveal that farmers using pesticides constitute 3 to 8 percent of all adopters of improved practices for certain crops, and 38 percent for the cotton crop (Rangaswamy, 1986: 481). Variations in pesticide use may be associated with the nature of farming and production inputs, economic and educational status of farmers, crops grown and extension agency impact and so on. We have not come across any study reporting on the farmer's pesticide preferences and usages.
7. *The Hindu,* November 26, 1986, p. 19
8. *Science Today,* September, 1986, p. 44.
9. *The Indian Express,* April 25, 1985.
10. *The Indian Express,* January 27, 1986, p. 8.

References

Bhalla, A.S. and D.D. James. "Technological Blending: Frontier Technology in Traditional Economic Sectors," *Journal of Economic Issues.* vol. XX, no. 2, June 1986, pp. 453-462.

Crump, Andy. "Pesticides Take Their Toll," *The Weekly Magazine of the Hindu,* December 7, 1986, p. 1.

Dover, Michael. "Getting Off The Pesticide Treadmill," *Technology Review,* Nov./Dec., 1985.

Girimaji, Pushpa. "Harmful Effects of Pesticide Residues," *The Indian Express,* June 18, 1985, p. 11.

Iyengar, K.R.S. "World Conference on Religion and Peace at Nairobi," *The Hindu,* August 21, 1984, p. 17.

Joshi, P.C. "Culture and Cultural Planning in India," Parts I and II, *Economic and Political Weekly,* vol. XVII, no. 50, December 10, 1983. pp. 2128-2131.
vol. XVIII, no. 51, December 17, 1983, pp. 2169-2174.

Kannan, A. and N.C. Ramanuja Chari "Science, Human Values and The Whole Man," *The Hindu,* December 30, 1986, p. 17.

Krishna Murty, C.R. "Are Chemical Pesticides Indispensable?" *Science Today,* September, 1983, pp. 15-16.

Mills, C.W. *The Sociological Imagination,* Oxford University Press, 1959, pp. 15-16.

Ramadass, V. "Kuttanad: Perils of Pesticide," *The Weekly Magazine of the Hindu,* June 15, 1986.

Ranga Swamy P. "Technology Policy for Dryland Agriculture: Some Issues and Approaches," *The Indian Journal of Agricultural Economics,* vol. 41, no. 4, October-December, 1986, pp. 476-486.

Sharma, C.B.S.R. "Instability in The agro-ecosystem Due to Pesticides," Paper presented at the 4th International Conference On the Perception and Management of Pests and Pesticides, Chiangmai, Thailand, January 7-11, 1985.

PART TWO

LABOUR MARKETS, AGRICULTURAL DEVELOPMENT AND TECHNOLOGICAL CHANGE

6

DECENTRALIZATION, INFORMALIZATION AND THE AGRICULTURAL LABOUR MARKET IN ADVANCED WESTERN SOCIETIES

ALLESANDRO BONANNO

Introduction

The paper addresses the issue of the relationship between the agricultural labour market and the phenomena of decentralization and informalization in rural areas of advanced Western societies. In the last few decades the agricultural labour market has been characterized by a surplus of labour. Despite a reduction in the size of the labour force employed in agriculture in both Europe and North America, agricultural operations need less than the available work force. This reduced demand has freed agricultural labour onto the market. Though the process of expulsion of labour from agriculture has been a characteristic of capitalist societies since the period of primitive accumulation, the phenomenon today is qualitatively different from before. In the past the expulsion of labour from agriculture meant a definitive passage of this labour to urban and/or manufacture activities. Today, on the contrary, a mixed social figure has emerged, that of a multiple job holder. This phenomena is associated with the growth of part-time and subsistence farming, as these activities are insufficient by themselves to support the farm family. In order to obtain complementary income farm family members offer their labour on the market. Simultaneously, in recent years farming has been subject to declining revenues, particularly in North America. The farm crisis has influenced the middle segment of the agricultural spectrum in particular, as small farm families can count on off-farm income and large farms still maintain their share of the market. The crisis of the middle sector of agriculture has freed additional labour onto the market.

The availability of labour, coupled with trends in urban areas which do not allow for rural-urban migration, provides a fertile ground for the development of forms of decentralization and informalization. Decentralization refers to the process of relocating industrial enterprises from traditional industrial areas to rural regions and/or the delegation of production to small family size enterprises. Informalization refers to the creation and maintenance of unorthodox structures of production such as cottage industry, house work, piece work, etc. Both decentralization and informalization can develop when cheap resources are available, particularly labour. The development of informal and decentralized activities has been regarded as a solution to economic stagnation. In this respect State involvement has been fundamental. However, the impact of decentralization and informalization on rural communities has not been conducive to development nor to the growth of external economies. Given the multinational and/or non-local nature of these enterprises, the income of local workers is often the only source available to stimulate the economy. Furthermore, the limited amount of wages paid depresses the aggregated demand and consequently hampers growth. The wealth generated is transferred outside the producing region so that while some accumulation takes place, socio-economic growth of the region is denied. At the local level the absence of economic growth generates stagnation which in turn involves the maintenance of precarious occupational structures. Labour is then forced to rely on part-time farming activities and low-paying off-farm jobs. Simultaneously, this situation fosters the persistence of a sizable yet economically precarious small-farm sector.

Selected Socio-economic Characteristics of Agriculture in Advanced Western Countries

The agriculture of advanced Western countries in the 1980s present characteristics that are significantly different from those of one or two decades ago. It is important, then, to mention briefly some of these characteristics, as they are relevant for the study of decentralization, informalization, informalication, and the labour market.

At the structural level, the agriculture of advanced Western countries is characterized by a concentration of land and sales. In the USA, for example, at the beginning of the decade five percent of all landowners controlled 48 percent of all land, while in Europe (EEC) 11 percent of all farms controlled 50 percent of all land. In Canada the total sales of large farms increased in the last decade from 32 to 56 percent of total agricultural sales, while the proportion of these farms out of all farms rose from 6 to 15 percent. Trends toward concentration have been paralleled by the persistence of a sizable small farm sector. In the member countries of the European Community 42 percent of all farms are five hectares or less in size and 60 percent are under 10 hectares, though they occupy only 13 percent of all available land. Similarly, in the United States, small farms with less than $10,000 in sales account for half of all farms in the country.

Differences exist between European and North American countries in terms of actual size of farms as well as in the manner in which farm size statistics are compiled. In the USA and Canada statistics categorize farms according to their acreage and their total sales. This second measure is usually considered more appropriate given the differences that the production of particular crops can make. In this respect, intensive cultivations, such as those of vegetables and fruits, generate much greater values per acre than intensive crops such as grains. In Europe it has been difficult to classify farms according to their sales so that acreage is the only measure available. Nevertheless, despite the existence of a nucleus of specialized farms that can generate high revenues employing a limited amount of land, there is a strong correlation between small farm size and low total sales (Bonanno, 1987).

The land concentration trend is also more rapid in North America than in Europe due to differences in land value (De Benedictis, 1980; Klatzman, 1978; Perez Blanco, 1983; USDA, 1984). The limited amount of agricultural land available in Europe has generated high land prices which are insensitive to changes in agricultural revenue. Moreover, the high potential for urban development has produced a *de facto* stagnation of the land market. In the USA and Canada, on the contrary, agricultural land values follow general changes in the prices of agricultural outputs. Accordingly, after a rise in the price of land during the late 1960s and early 1970s, there has been a constant decrease. This decrease in land value has generated an impetus for the concentration of land, as large farms can provide more collateral for new acquisitions and new entrances have been discouraged. Simultaneously, owners of smaller farms have seen the value of their collateral reduced and have become engaged in long-term debts which often result in foreclosures.

A common characteristic of both North American and European countries is the tendencial polarization of the agricultural structure. The growth of the large-farm sector and the persistence of the small-farm sector has created two completely opposite "agricultures" which confront different problems and alternatives. As far as middle-sized farms are concerned, though they still represent an important component of agriculture in all advanced Western countries, their overall position can be effectively described as one of crisis in North America and stagnation in Europe.

Agriculture in advanced Western countries employs fewer individuals in the 1980s than one or two decades ago, both in absolute and relative terms. Sharp reductions in the labour force employed occurred particularly in Europe where agricultural labour was a fundamental component of the total domestic employment of many countries. In Italy, for example, agricultural labour accounts for 14 percent of the total labour force, while in the 1960s this figure was 30 percent. Similarly, there has been a decline in France from 22.3 percent in 1960 to 8.9 percent in 1980; in West Germany from 31.2 to 6.2. In the United States and Canada the reduction of agricultural labour has not been as sharp as that of Europe, given the already limited proportion of farm employment out of the total domestic employment. Nevertheless, in the USA the percentage of labour engaged in agricultural activities decreased from 8 in the early sixties to 3.5 in 1985, in Canada from 12.6 to 5.6 in the same period.

The reduction in agricultural labour is the outcome of complex historical trends. In brief it can be stated that such a reduction is related (a) to the application of labour-displacing technologies in the farming process, (b) to the accumulation crisis experienced in the sector, and (c) to ideological and cultural factors that have motivated farm residents to abandon farming.

Despite this loss of employment, agriculture today still supplies labour to other economic sectors. In the 80's, however, this phenomena is qualitatively different from before. Historically, agriculture represented a reservoir of labour for the expansion of manufacture first and the tertiary sector afterward. Loss of agricultural labour meant a well-defined passage from the status of agricultural worker to that of urban worker. In recent years the separation between work in agriculture and in other economic sectors has become increasingly tenuous for many rural residents, resulting in the emergence of a new social figure of the multiple job holder.

In order to illustrate the qualitative changes that have occurred in the agricultural labour force, the case of family farm labour will be employed. This selection does not imply a dismissal of the importance of the role that other social figures, such as farm workers, play in the agricultural labour market. Rather, it purports to be a simplified analysis of a complex and contradictory social process.

Labour, Agricultural Crisis, Decentralization and Informalization

The consequences of the crisis of accumulation experienced in agriculture in the past few years, the current trends in agricultural labour and land markets and new forms of organization of production are central to the present discussion. It is important, then, to underscore briefly the basic characteristics that pertain to the relationships among these phenomena.

In recent years agriculture in advanced western countries has been characterized by a serious economic crisis. The effects of this crisis, however, have not affected all of the various types of farms nor the various agricultural social figures in the same manner. As has been very well documented for the case of North America (USDA, 1985) and Europe (OECD, 1984), the crisis concerns primarily the middle sector of agriculture, leaving untouched the large-farm sector which continues to expand and grow. Small farms are also affected by the crisis. However, in their case alternative sources of employment and income available to farm family members through part-time farming have eased the impact of the crisis. Given the limited dependency of small-farm families on agricultural income (Arkleton, 1983) it has been argued that the persistence of small farms is now a fundamental characteristic of advanced agriculture (Bonanno, 1987).

Part-time farming and the crisis of full-time farms has freed a portion of the labour force once involved exclusively in agriculture for off-farm activities. Furthermore, an additional portion of labour has been freed by the crisis of agricultural communities, which is consequential to the farm crisis. The availability of this labour is one of the major characteristics of present rural areas both in North America and Europe, for it is offered on the market at lower prices than its urban counterpart. First, local labour demand is low, as rural areas have experienced lower levels of economic expansion, and employment growth has been inadequate to absorb increases in the labour force resulting from population growth (USDA, 1984; OECD, 1984).

Second, lower off-farm wages can be compensated for with benefits derived from farm activities in the form of both commercial part-time farms and subsistence farms. Finally, the act of complementing farm income with off-farm earnings extends its influence from the sphere of production to others external to it, such as labour disputes and labour unionization (Bonanno, 1987).

Historically, lower wages have been a fundamental source of labour unrest and a major stimulus for union organization (Piven and Cloward, 1977). This motivation factor is largely devalued in the case of part-time farming, for additional income and/or non-monetary benefits are sought in the expansion of the farming activity. Traditional industrial relations are thus modified so that structural contradictions are resolved at the individual level. Similarly, the implementation of individual options in the working sphere decreases the potential for unionization and for an increase in labour costs.

In North America inexpensive land has played an important role in the present situation of rural areas. Declining and/or low land value generated a devaluation of the farmer's assets, exposing him to financial difficulties and providing a stimulus for the non-farm use of land. Cheap land attracts investors, such as industrial firms, which see in the lower prices a manner to curb costs of production and enhance their potential to compete on the market. It is not a coincidence that low land costs in the USA have been identified as one on the major factors influencing rural industrialization (Chantfort, 1983; Martinez, 1985).

The combination of cheap resources, stagnating economic conditions, and availability of cheap labour has fostered the development of two particular forms of organization of production: decentralization and informalization.

Decentralization refers to the process of decentralizing industrial activities from traditional, old industrial areas to rural regions and/or the delegation of production from centralized traditional units to decentralized small firms. In the case of North America it means the relocation of industrial activities to rural areas in primarily Southern and Western regions (Chantford, 1983; Horan and Tolbert, 1984; Martinez, 1985; Swanson and Skees, 1985). Increase in production costs and the strength of labour in the core sector has provided ample support for such strategies. Accordingly, "Industry has been rapidly decentralizing over the past two decades" (Martinez, 1985).

The industries involved are generally low-wage, labour-intensive manufacturing establishments such as textile mills and plants producing clothing, furniture, leather goods, rubber and plastic products and, more recently, automobiles. Such industries can greatly profit from the cheap labour and low costs of production available in rural areas. In the United States these establishments, tend to be branches of corporate firms, and a large segment (80 percent) of rural employment tends to be provided by establishments with more than 100 employees (Chantfort, 1983; Martinez, 1985). Furthermore, in many cases plants in rural areas tend to be larger than similar ones found in old industrial regions.

"One factor (for the occurrence of this phenomenon) is lower nonmetro land costs, which encourage sprawling complexes employing many workers. Another is that large, mature firms often choose to settle in nonmetro areas, where labour costs are lower. Also, some nonmetro establishments are large simply because they are not surrounded by specialized suppliers, but have to produce many of their own inputs" (Chantfort, 1983: 9).

Despite the creation of new jobs, the decentralization process has been associated with rather unstable employment. These jobs are often exposed to foreign competition, as non-domestic producers can count on even cheaper costs of production.

"In the past, these industries have been shifting from metro to nonmetro areas, providing an important source of employment growth in rural and small-town America. Now, however, an increasing shift of these operations overseas jeopardizes this source of jobs. Because the low-wage, labour-intensive industries employ a relatively large proportion of the local labour force in some rural communities, the loss of such establishment can hit these communities especially hard" (Chantfort, 1983: 9).

In Europe the process of decentralization differs from that of the USA, for it is characterized by the delegation of production process to smaller units. As has been documented for the case of Italy (Bagnasco, 1981; Mingione, 1985, 1986; Pugliese, 1983), the firms involved in decentralization are rarely branches of larger corporations. On the contrary, they are small and/or family-owned businesses which contract tasks from large operations. Their markets of operation both in terms of inputs and outputs are controlled by the larger corporation which operates as a monopsony (the only supplier) and a monopoly (the only buyer) for their products. However, the European and the American cases are similar when costs and skills of labour are compared. In both cases low costs and unskilled labour are stable characteristics.

Informalization is the other process that has played an important role in the reorganization of production in rural areas. The concept of informal economy has been employed to describe a diverse set of labour activities which have been interpreted in conflicting manners. In the present context it refers to labour activities which exclude house chores (self-work) and formal labour activities regulated by legal contracts and/or trade union agreements. Informal activities define, then, a large gray area which includes the creation and/or maintenance of unorthodox structures of production such as cottage industry, housework, piece work, etc., with the use of marginal labour often paid below regular (legal or market) fares and with no fringe benefits and/or employment contract. Informal activities may also involve the extension of domestic work outside the domestic sphere to include forms of exchange of performance and/or non-monetary trade and the generation of services.

Agricultural Crisis, the State and Growth with Underdevelopment

The phenomena of decentralization, informalization, agricultural crisis, and the overall restructuring of the production system are connected with the role of the State. As has been indicated elsewhere (Bonanno, 1988), the State's role in advanced capitalist societies has been studied and interpreted in differing manners. Yet there is general agreement, at least in the progressive camp, with respect to its limitations and contradictions.

The role of the State in advanced Western societies since the end of World War II has been that of general mediator between conflicting interests of the major social classes. This role of "pacifier", as Offe (1984) has illustrated, consists first (but not primarily) of the support of important needs of the working class. Due to an increase in the power of the working class (Korpi, 1983), the State provides social service assistance to its members even in the case of those who are marginalized by the competition present in a market society (unemployment, low wages, etc.). Simultaneously, the state regulates disputes between the working class and the capitalist class through legislation and direct arbitration. This normative role of the State involves the legitimation of the role of trade unions (and other organizations of the working class, such as political parties) in society and, through this, the limitation of the range of action available to the working class against the bourgeoisie. Second, the State guarantees the possibility of accumulation and, consequently, supports the interests of the bourgeoisie. However, as indicated in many recent works (Block, 1977; Offe, 1984: Poulantzas, 1978) these interests do not correspond to the short-term needs of any specific fraction of the bourgeoisie itself, but represent the interests of the class as a whole. The State is thus supportive of the general conditions that allow long-term accumulation in society. From this point of view it is possible to argue that the position of the State as mediator is not a neutral position, but a class position. It favours the rule and the leadership of the bourgeoisie and the continuous existence of the present relations of production (Block, 1977; O'Connor, 1973; Offe, 1984; Poulantzas, 1978).

During the course of the last decade the role of the State as "pacifier" became increasingly untenable, as fundamental contradictions emerged in mature capitalist societies. As has been documented in a number of cases (Castells, 1980; Offe, 1984; Poulantzas, 1978), the State is unable to solve contradictions related to the lack of accumulation (regional underdevelopment) and social problems in large areas of advanced societies. This inability of the State to address the needs of the working class and satisfy the bourgeoisie's interests in accumulation exposes the State itself to a mounting legitimation crisis to which it is unable to respond without compromising possibilities for further accumulation. In this respect, the accumulation crisis is accompanied by a legitimation crisis that generate tension between the State and various segments of the bourgeoisie and between the State and the working class. More importantly, the legitimation crisis limits the autonomy of the State and its range of action, as its potential for acquiring financial resources is eroded, (Offe, 1972).

Accumulation and legitimation crisis and the action of the State to avoid them are integral parts of the farm crisis and the phenomena of decentralization and informalization. In the case of decentralization, the creation of non-farming jobs for rural areas becomes a historically fundamental feature of State action in order to avoid crises of accumulation and legitimation. In other words, in its attempt to enhance accumulation and legitimation, the State is willing to support the creation of business in relatively marginal areas so that the reproduction of capital, the generation of consensus, and social stability are obtained.

The relationship between the State and the creation of off-farm jobs in rural areas is twofold. First, the State is willing to support the interests of corporations to invest in regions that are characterized by lower costs of labour, land, and raw materials. Furthermore, the State is willing to guarantee discounts in the construction of the plant and the development of infrastructures. Recent cases in point in the United States are the incentives given to private capital by the state governments of Kentucky and Tennessee for the construction of Toyota and Saturn plants, respectively. Incentives have also been given for the education of the labour force in these regions. The result of these operations, according to the officialdom of the State, would be that of supplying a number of jobs to curb local unemployment and stimulate local demand so that the total process of accumulation would be reestablished. Second, at the level of legitimation, the creation of new jobs enhances the ability of State officials to maintain support among their constituents and ensure social stability.

In the process of attracting new jobs to rural areas the State is willing to settle for lower paying and often part-time, temporary employment to replace higher paying jobs lost in the past, or simply to curb unemployment. However, this process creates contradictions at the level of both accumulation and legitimation. As far as the former is concerned, low paying, part-time jobs have traditionally reduced the capacity to generate external economies and, consequently, to trigger further socio-economic development. Given the multinational and/or non-local nature of these corporations, the income of the local workers is often the only source available to achieve this goal (Summers 1986). Furthermore, the decreased aggregated demand generates a negative effect on the overall process of accumulation (Mingione, 1985). At the legitimative level, the expectations of the unemployed workers may not be met by these kinds of new job, and dissatisfaction among local residents may be generated.

This is not to say that the establishment of new plants in rural areas does not generate employment. Rather, this employment is precarious in nature and historically has not been conducive to stable socio-economic development. The strategy of the "Poles of development" experienced in the 1960's and 1970's both in Europe and in the USA is a case in point.

The agricultural crisis and the processes of decentralization and informalization do not allow for the establishment of traditional forms of labour capital relations. The conditions for the existence of industrial activities remain those of cheap means of production, cheap labour and constant assistance on the part of the State. Once these conditions are violated and there is an increase in the cost of labour and/or costs of production, the corporations involved either move their plants elsewhere or judge them unprofitable and dispose of them through selling or closing. In the case of contracting, given the high supply of labour, new contractors are available in the formal or informal market. Furthermore, any of these operations will put a strain on the fiscal crisis of the State as it is called upon for further economic intervention. The State strategy to induce private accumulation to solve social problems becomes the State's worse accumulation and legitimation crisis.

Alternatives to the present crisis of accumulation and legitimation appear problematic. The solution to enhance accumulation through decentralization and informalization is contradictory. More specifically, it is based on the assumption that conditions of underdevelopment must be maintained in the region in question. The limitation of the aggregated demand and the impossibility of developing external economies reduce the potential for accumulation. Furthermore, the commodities produced through decentralized and informal activities need to be sold in external markets which tend to be saturated and, consequently, inadequate to absorb the available supply. In this respect the short-term solution to the reduction of the process of accumulation becomes the first step in a long-term stagnation.

At the regional level the notion of *Growth with Underdevelopment* seems to exemplify the situation best. The creation of wealth in loci is transferred for the most part to other regions. Low wages are the only compensation to this process for the local population. Thus, though accumulation takes place in these areas in the short run, its beneficial effects on the local working class are largely eliminated. In the long run the effects of a shrinking aggregated demand will be damaging for the entrepreneur, and the advantage of employing cheap resources and labour will be nullified by a lack of markets. In this respect the initial stagnation of local economies will be replaced by rampant economic crisis.

Despite these problems the present socio-economic conditions in advanced Western countries foster the development of decentralized and informal activities. Given the diminishing supply of labour needed in agriculture and the crisis of industrial establishments, workers are freed on the market for decentralized and informal activities. As these jobs, however, are precarious in nature, alternative sources of income are needed. In this respect welfare programmes and subsistence farming may continue to play an important role in the overall socio-economic structure of rural regions.

Note

The inverse is also true in that ideological and cultural factors are responsible for urban-rural migration and the persistence in agriculture of many farm families despite difficult economic conditions.

References

Arkleton. "Part-time Farming in the Rural Development of Industrialized Countries." Report of a Seminar Held in Scotland from October 16 to 21, 1983. The Arkleton Trust.

Bagnasco, A. "La Questione Dell, Economia Informale." *Stato e Mercato*,(1 April 1981): 173-196.

Block, F. "The Ruling Class does not Rule." *Socialist Revolution* (3), 1977, 6-28.

Bonanno, A. *Persistence with Legitimation; The Persistence of Small Farms in Advanced Western Societies*, Boulder, Co :Westview, 1987.

Bonanno, A. "The State and Land Reform : The Case of Italy (1944-1962)," *The Sociological Quarterly*, (i), 1988, 131-147.

Castells, M. "Toward the Informal City? High Technology, Economic Change, and Spatial Structure." Working Paper No. 430, Institute of Urban and Regional Development, University of California Berkeley, 1985.

Chantfort, E.V. "Will Midsize Farms Fade as Small Farms and Big Farms Multiply." *Farmline* 3, (April 1982): 4-6.

De Benedicts, M. "Sviluppo e Ristagno Dell'Agricoltura nel Mezzogiorno." In M. De Benedictis (ed.) *L'Agricoltura Nello Sviluppo del Mezzogiorno*. Bologna : Il Mulino, 1980.

Horan, P. and C. Tolbert. *The Organization of Work in Rural and Urban Labour Markets*, Boulder, Co : Westview Press, 1984.

Klatzman, J. *L'Agriculture Francaise*. Paris: Edition du Sevile, 1978.

Korpi, N. *The Democratic Class Struggle*. Boston : Routlege and Keagan, 1983.

Martinez, D. "The Bottom has Faded, and Many Rural Areas Lag Behind." *Farmline* 2 (April, 1985,): 12-14.

Mingione, E. "Informalization and Households' Survival Strategies in Industrialized Societies". In M.P. Smith and J.R. Feagin, *The Capitalist City*, Oxford : Blackwell, 1986.

Mingione, E. *L'Approccio Sociologico e le Economie Urbane di Sussistenza*. University of Messina. Messina Italy, 1985.

O'Connor, J. *The Fiscal Crisis of the State*. New York: St. Martin's Press, 1973.

OECD. *Issues and Challenges For OECD Agriculture in the 80s*. Paris: OECD,1984.

Offe, C. *Contradictions of the Welfare State*. John Keane (ed.) Cambridge Mass. : MIT Press, 1984.

Offe, C. "Advanced Capitalism and the Welfare State." *Politics and Society*, 1972.: 479-488.

Perez Blanco, J.M. "Rasgos Macroeconomicos Basicos de la Evolucion de la Agricoltura Espagnola 1964—1982." In *La Nueva Agricultura Espagnola*, Madrid :PDEE, 1983.

Piven, F. and R. Cloward *Poor People's Movements Why They Succeed How They Fail*. New York : Pantheon Books, 1977.

Poulantzas, N. *L'Etat, Le Poivoir, Le Socialism*. Paris : Presses Universitaries de France, 1978.

Pugliese, E. *I Braccianti Agricoli in Italia*. Milan : Franco Angeli, 1983.

Summers, G.F. "Rural Industrial Ratios," *The Rural Sociologist*, Vol. 3, (May 1986): 181-187.

Swanson, L. and J. Skees *Public Policy for Farm Structure and Rural Well-Being in the South*. OTA Report. Washington, D.C, 1985.

USDA. The Current Financial Conditions of Farmers and Farm Lenders. USDA, ERS, *Agricultural Information Bulletin*, no. 490, 1985.

USDA. Handbook of Agricultural Charts. *USDA Agricultural Handbook*, no. 637, 1984.

7

MARKETS AND TECHNICAL CHANGE
The Case of the Portuguese Dairy Sector

MANUEL BELO MOREIRA

Introduction

On the basis of a case study – the dairy sector in the minifundia of Northern and Central Coastal Portugal – I intend to illustrate how, despite the rigidity of land tenure structures, new market forces have contributed to the transformation of traditional farming systems. This new dynamism is the result of the deep changes that affected Portuguese economy and society at the end of the 1959's and of the prominent role played by the State.

I will focus my attention on the adoption and diffusion of mechanical milking and bulk collection of milk and, above all, on the "original" way this technical change was carried out. Essentially, it took a different form from that experienced in other countries, and in this particular Portuguese region, the diffusion of innovations did not force small and very small producers out of milk production.

Table 7.1 — **Structure of Milk Production**

		nº of Producers with 1-5 Cows	% of total nº of Cows[a]	nº of Producers Total	nº of Cows Total
1967	Minifundi[b]	43,457	90.1-95.3	43,839	66,051
	Mainland Portugal	68,670	60.4-80.8	71,241	135,909
1981	Minifundia	59,009	55.1-76.1	63,045	146,448
	Mainland Portugal	89,060	40.8-57.8	98,670	265,777

[a]. This is a theoretical margin showing a maximum and minimum.
[b]. Distritos: Aveiro, Braga, Coimbra, Porto e Viana do Castelo.

It is my aim to demonstrate that this "originality" can be explained through a consideration of the special features characterizing the socio-structural conditions of the region in question and by the role of the State, particularly in regard to the creation after 1967 of collective centres for mechanical milking.

Some of the Principal Features Contributing to the Transformation of Minifundia Agriculture in Northern and Central Coastal Portugal.

The situation at the end of the 1950's

At that time, agricultural production was spatially divided into two areas. In the Northern part of the country, where most of the farms were small, very small, and fragmented into separate plots of land, family farming prevailed and involved large areas of rented land and sharecropping. In the South, capitalist and extensive latifundia prevailed until the Agrarian Reform was implemented in 1974.[1]

The subjects of this study are overpopulated minifundia in the coastal strip of Northern and Central Portugal (districts of Aveiro, Braga, Coimbra, Porto, and Viana do Castelo) which were characterized by the following features:

Family farms were devoted primarily to the cultivation of mixed crops and animal breeding. Besides dairy and meat production, bovines also supplied the labour and organic fertilizer essential for intensive agriculture.

Labour was abundant allowing high levels of intensification although living conditions, at both the economic and social levels, were often very low.

Farms were characterized by very low levels of use of industrial inputs, which also generated low levels of productivity. Labour, soil and cattle were underutilized.

During the fifties, prices of the principal agricultural products were kept constant in accordance with a governmental effort to supply inexpensive labour to industry. This policy was combined with an absence of free unions and with the political repression of workers movements. Constant agricultural prices, rising costs of industrial inputs, and a fragmented agricultural structure made farming a very unattractive type of investment.

By the end of the fifties, Portuguese economy and society ended their isolation and initiated a process of opening to the exterior which had obvious repercussions on the agricultural sector. This process was primarily caused by two different phenomena. The first of these was the economic expansion of the rest of Europe, which created new job opportunities and stimulated the out-migration of labour from Portugal. The second cause was the outbreak of a colonial war which was to last over 15 years and which mobilized 150,000 able-bodied men at one time (about five per cent of the working population in 1970).

Job opportunities abroad and the colonial army demand for men motivated many dissatisfied rural workers to join the rural and agricultural exodus. Simultaneously, industrial and urban development and the growth of tourism stimulated changes in food consumption patterns.

The Role of the State

During the first half of the 1960's the impossibility of maintaining constant food prices[2] and the end of State intervention concerning the land tenure structure created the conditions for the development of an agricultural policy strongly devoted to the development of dairy and livestock production. It combined credit, guaranteed prices, and subsidies to be granted to agricultural producers and in some cases to processing firms. In short, it aimed at stimulating livestock production with the dual objectives of contributing to the satisfaction of the country's food needs and creating new production opportunities for small farms.

To achieve this objective the adopted policy proposed a technical model of livestock development which followed almost mechanically those adopted in industrialized countries. This model was based on intensive livestock breeding, the use of highly specialized and highly productive animals, industrial inputs such as compound feeds,[3] and sophisticated equipment such as free or semi-free stable premises equipped with mechanical milking and bulk tanks. This technical model was implementend with the assumption that high increases in productivity and substantial improvements in milk hygiene would be achieved. Furthermore, it was designed exclusively for large farms in the capitalist region (see table 7.2) where investments were.

Table 7.2 - **Structure of Holdings**

Classes of Area	Mainland Portugal				Medium and Large Holdings Zone[a]				Dairy Minifudio Zone[b]			
ha	nº	%	Area	%	nº	%	Area	%	nº	%	Area	%
landless	4131	.5	1058	-	2769	1.6	557	-	381	.1	45	-
<1	344255	43.9	209867	4.1	60671	35.6	33869	1.2	157288	54.8	87625	12.6
1-4	298892	38.1	663713	12.8	61986	36.4	137443	4.8	102347	35.7	208693	30.0
4-20	116113	14.8	1004860	19.4	32456	19.1	283238	10.0	25082	8.7	187014	26.9
20-50	12068	1.5	405124	7.8	6171	3.6	201447	7.1	1267	.4	39330	5.7
50-100	3587	.5	283427	5.5	2418	1.4	180191	6.3	261	.1	19679	2.8
100-500	3941	.5	990425	19.1	3100	1.8	735818	25.9	174	.1	60829	8.8
>500	957	.1	1624428	31.3	743	4.4	1267509	44.6	39		91518	13.2
Total	783944	100.0	5182902	100.0	170314	100.0	2840072	100.0	286839	100.0	694733	100.0

[a]Zones: Beja, Evora, Lisboa, Portalegre, Santarem, and Setubal.
[b]Zones: Aveiro, Braga, Coimbra, Porto, and Viana do Castelo.

considered to be profitable. The minimum size proposed for dairy production was then fixed at around 30 cows. No policy was specially designed for small farms despite the fact that minifundia had a strong tradition in dairy and livestock production.

In that period it was often suggested and hoped that migration would free some land that could then be incorporated to form larger holdings. However, this hope did not become a reality and the goal was never achieved. On the contrary, rather than promoting land tenure reorganization, migration contributed to its stagnation and even to its further fragmentation.

As far as milk production was concerned, hope had far exceeded reality, for only a small number of medium and large farms adopted the proposed model, and contrary to the desired outcome, it was mainly in the minifundia that dairy farms developed (see table 7.1).

Here local farmers organizations (both cooperatives and corporative organizations) encouraged small and very small farmers to produce milk and take advantage of the incentives intended for medium and large-scale farms.

The larger role played by small farms in milk production was eventually recognized by the government, which initiated and promoted production among these farms. The reason for this change in government policy was twofold. On the one hand, it was due to the fact that small and very small farms were responsible for supplying most of the milk to the dairy industry, and on the other, it was related to the negative impact that the policy could have had on employment in the small farm sector.

Transformation of the traditional farming system of the minifundio[4]

Transformation in the minifundio traditional farming system during the last quarter of the century were slowly carried out without major disruptions[5] as farmers were able to adapt the traditionally employed farming system to the newly created situation. Following are some of the principal factors which influenced such an adaptation.

A Rigid Structural Framework

Small fragmented farms characterized the Northern and Central part of the country. Attempts to increase the average size of farms through land purchase failed[6] due to a high demand for land stemming from a number of sources including farmers, rurally located industries, residential developers, etc. Consequently, the existing demographic pressure on land contributed to the worsening of the structure of the minifundio.[7] Particularly serious was the condition of the market for small plots of land, as they were in great demand by returning immigrants endowed with savings to invest.

A possible alternative was that of increasing the portion of rented land. However, various factors have hindered such a possibility. On the one hand, absentee small land owners (immigrants) were reluctant to rent their land as they preferred to have it available at the time of their return from abroad. On the other hand, legal procedures and political postures prevalent at the time discouraged rental of land.

Finally, the outflow of rural labour was halted in the 1970's as job opportunities in other economic sectors ceased to exist in the same quantity as in the 1960's. This situation was further aggravated by the sudden return of about one million former colonists following the independence of the African territories. It is evident that, except for a few occupational outlets, agricultural labour did not have sufficient opportunities to escape an employment crisis.

These aspects, along with the impossibility of financing plans aimed at reducing the number of farmers, such as an anticipated retirement scheme[8], have created an extremely rigid agricultural structure.

Changes in Traditional Farming Systems

Due to the changes which occurred in the socio-economic sphere, minifundio holdings were forced to a progressive integration into the international market.

As far as the labour structure was concerned, migration patterns and the limited demand for stable employment contributed to the strengthening of a multi-activity network characterized by farm and off-farm employment.

At the economic level, international integration represented a stimulus for the industrial and urban development of the country. The agricultural policy was also affected as measures such as price subsidies, credit policies, policies for the development of mechanization, soil conservation, improvement of seeds, and artificial insemination were introduced.[9]

The changes in the socio-economic conditions and the expansion of consumption with the parallel creation of new "needs" forced small farm families to abandon, progressively, self-sufficiency and become more dependent on monetary relationships and off-farm activities. This process implied changes in the farming system. Such changes were introduced through various processes of adaptation capable of responding to both macro-economic domestic conditions and to the diversified international situation.

In essence, two distinct patterns of adaptation can be identified. First, farms located in the best edapho-climatic areas, well situated in regard to markets, benefiting from a good distribution and input supply network, and in a good position to profit from incentives provided by agricultural policies initiated[10] processes of intensification of production. More specifically, greater capitalization, the substitution of labour for capital, changes in methods of cultivation, a greater use of machines, and a greater specialization began to be carried out on these farms.

Second, farms situated in economically depressed areas and/or managed by aging and sometimes ill-prepared farmers initiated processes of progressive extensification, of increasing fallow ground, of using less intensive crop rotations and sometimes abandoning the worst plots located furthest from the farm house (Caldas, 1981). These processes of extensification were possible, however, due to the remarkable increase in off-farm revenues.[11]

The Growing Importance of Milk Production in the Minifundio of Northern and Central Coastal Portugal

Milk production in the minifundio of Northern and Central Coastal Portugal began to gain importance in the 1930's after the first dairy farms were established.[12] However, it was during the 1960's that productivity and production reached high levels. This phenomenon was related to the implementation of policies for the development of dairy farms. More specifically, it was an outcome of the policy of substituting indigenous cows with imported species along with improvements in the genetic potential of existing cows achieved through increasing artificial insemination and better feeding. From that time on, the number of producers increased (+43.8 percent between 1967 and 1981) and more cows (+121.7 percent) were added to the herds. This development in milk production took place primarily in the areas where a network for milk collection was established through programmes of collective collection carried out by organization of producers. Before 1974, they were corporative organizations, and after 1974 they were part of cooperatives.[13]

Profiting from the structure of collection, farming organizations decided to experiment with a model of milk production suitable for small fragmented holdings. Accordingly, in 1967 one cooperative together with the corporative organization began extending to small and very small farms financial and other benefits aimed at improving mechanical milking and bulk collection. This action resulted in the emergence and later development of collective mechanical milking centers (Moreira, 1982 and 1984).

Mechanical milking was adopted first by the collecting organizations that could obtain better quality milk. They could count on monetary subsidies granted for the installation of mechanical milking and bulk tanks, which for some years covered as much as 80 percent of total equipment costs.[14] Furthermore, mechanical milking was attractive for those farmers benefiting from guaranteed sale of their milk.

The policy of guaranteed sales was fundamental in the development of a network of collective mechanical milking centers in the minifundio area. It also contributed to make milk production a more attractive business to the extent that between 1967 and 1981, the number of farms increased by 19,206 units, representing 70 percent of the increase in the total number of farms. The number of cows also grew to 80,397 units, representing 61.9 percent of the increase in the total number of dairy cows.

Technical Changes in the Minifundio Dairy Farming System

Unitil the 1950's, with the exception of a few less significant examples, the dairy industry in the minifundio was a minor component of the traditional farming system. Family farms practiced mixed farming systems featuring bovines supplying traction power and organic fertilizer for vegetable crops. Beef and milk constituted secondary outputs with milk seldom exceeding 20 percent of the total production (Rodrigues, 1951).

Despite the importance of bovines in this farming system, the relative insignificance of milk production was instrumental in the creation of a situation in which deficient bovine maintenance was the norm. It was characterized by the very irregular feeding of cows, a lack of adequate animal diets, and poor hygiene. This situation together with the predominance of indigenous cows resulted in very low levels of milk production per cow.

Integration into the international market forced these farms to adapt to a diminishing labour force, which in turn led to the emergence of labour saving techniques. Such techniques consisted of the simplification of cultivation strategies[15] and the increased use of mechanical equipment.[16]

Increases in the level of consumption of milk and dairy products, of which Portugal had almost always had a deficit, and policies for the development of production also made the dairy industry an increasingly attractive sector. This signified that farmers were increasingly able to purchase animals with a more productive genetic potential[17] which were also immediately capable of better responses to the richer and more balanced diet available.

Furthermore, higher production and productivity were achieved as cows were decreasingly employed in heavy work such as ploughing, now almost exclusively executed by machines.[18]

From the producers' point of view, mechanical milking in collective centres generated various advantages. First, the simple fact of having a collective milking centre equipped by the collecting organizations was a guarantee of sale at the best price without having to invest. Second, mechanical milking was advantageous as it reduced the hardship of milking, particularly benefiting aging farmers. Finally, it allowed better management of family work, as tasks could be distributed among the various family members.

In short, milking centres made work easier for farmers, allowed better management, created enormous advantages particularly in the cases of farm families with multiple job activities and, above all, benefited farmers without requiring direct economic costs. Among the disadvantages were the difficulties arising from having to lead cows to the milking centres. This task is particularly hard when there are three or more cows, when the distance they have to travel increases, and when the roads have too much traffic. Also, the outbreak of epizoa has been common in these situations, creating additional problems.

From the point of view of collective organizations, mechanical milking centres represented the fulfilment of their extension duties. By allowing small farms to participate in this process, their mandate to benefit the farmers they representated was accomplished. Furthermore, by building collective mechanical milking centres, collecting organizations improved milk quality and hygiene. The capacity to obtain larger quantities of milk in good hygienic and sanitary conditions enabled them to compete within the drinking milk market where milk is sold at higher prices and is paid for promptly without delays.[19]

As for refrigeration, the diffusion of bulk collection was much slower. This phenomenon was due to the fact that collecting organizations, which were responsible for purchasing equipment, were in charge of installation only when the collecting circuit serving such centres was ready to be totally refrigerated.

Finally, it is relevant to mention some figures which illustrate the evolution of these innovations in the minifundio area. For this purpose data supplied by the two most important Cooperative Unions, the Uniao de Cooperatives Leiteiras de Entre Douro e Mondego LACTICOOP and the Uniao de Cooperatives Leiteiras de Entre Douro e Minho-AGROS are used. At the beginning of the 1970's the quantity of milk mechanically processed and refrigerated was negligible in both of these Cooperative unions. In a period of about 10 years the development experienced was remarkable. In the LACTICOOP area in 1982 the number of existing collective centres was 573, of which 549 had bulk tanks, while the number of farms provided with individual mechanical milking was 430, of which only 225 were equipped with bulk tanks. The volume handled was also significant. In collective milking centres provided with bulk tanks 124,618 thousand litres were collected, compared with only 14,554 thousand collected by farms outside the cooperative. However, in the AGROS area the individually operated farms had higher productivity although their production was much below that of the collective centres. In 1983, there were estimated to be 508 collective centres with a production of 83,914 thousand litres, mostly refrigerated. There were also 1,293 individually operated farms which used mechanical milking techniques. Among these farms only 580 were equipped with bulk tanks and they produced 29,645 thousand litres, of which 16,808 thousand litres were of non-refrigerated milk.

Conclusions

In essence, changes in the farming system of the minifundio in Northern and Central Coastal Portugal indicate that, under certain circumstances, State intervention in defence of small farms resulted in benefits that can be considered of value not only to producers but also to the entire society. The adoption of dairy farming systems involving mechanical milking and bulk collection and put into practice through an "original" model of installing collective mechanical milking centres is a case in point. State intervention produced a number of important results. First, it stimulated both production and productivity while maintaining the minifundio structure. Second, it improved the living and working conditions of farmers. Finally, it contributed to a reduction in the agrarian and rural exoduses, which in a situation of generalized crisis and unemployment would have certainly aggravated social tensions and increased related social costs.

Notes

1. The principal family farming area was located in the Northern districts (Viana, do Castelo, Braga, Porto, Aveiro, Coimbra, Leiria, Vila Real, Braganca, Viseu, and Guarda and Faro). The capitalist and latifundia areas were located in the districts of Beja, Evora and Portalegre. The remaining districts situated in the border between North and Alentejo were also characterized by minifundia.
2. A strong crisis was then felt largely due to the escalation of the agricultural exodus. It produced not only an increase in wages but also a compulsory transformation of traditional farming systems.
3. With the encouragement given to the consumption of compound animal feeds, a new industry, practically non-existent in 1960, was created and has since then experienced a spectacular development.
4. This analysis was inspired by the works of Prof. Petit's team of INRA/ENNSA - DIJON.
5. No attempts to carry out structural reforms were made. This was true also for the period following the April revolution of 1974. Though an agrarian reform was eventually carried out, it involved only latifundia in the south.
6. Prices of agricultural land very often reach values very close to those demanded for urban land. In Northern and Central coastal areas of the country values as high as 2,600 - 4,000 US $/acre are common, sometimes going as high as 5,000 US $/acre and over.
7. In the minifundio districts of Central and Northern Coastal Portugal small and very small holdings increased their proportion out of the total number of farmers from 74.5 percent in 1968 to 81.4 percent in 1979. In Central and Northern inland areas they increased during the same period from 53.3 percent to 64.8 percent (Cordovil, 1984).
8. From 1977 onwards, the economic crisis and the country's foreign debt have led to very restrictive policies partly imposed by negotiations with the IMF (International Monetary Fund), which have not allowed any sort of serious attempt to implement such a retirement policy.
9. The industrial sector has experienced an enormous growth since 1960, characterized by a significant increase in the quantity of raw materials imported.
10. In Portugal these have only very recently been directed toward economically depressed areas.
11. In the Northern and Central districts, on average more than 30 percent of the family income came, according to a 1979 estimate, from emigrants' remittances, pensions and interest on savings (Cordovil et all.)
12. Some of these farmers have recently established farms in the Azores where edapho-climatic conditions are excellent for milk production.

13. Prior to 1974 some cooperatives ran a network of collection under the supervision of corporative organizations. After April 25, 1974 with the dismantling of the corporate organization, cooperatives gained the legal monopoly which they maintained until the recent entrance of Portugal into the E.E.C.
14. Between 1967 and 1981, 1420 collective milking centres were installed, of which 1152 were equipped with bulk tanks. This figure is much higher than that of private farms, equipped with mechanical milking and refrigeration (data obtained in the Direccao-Ceral de Pecuaria).
15. See the selection of texts edited by Lacticoop, 1983.
16. Between 1974 and 1983 the number of tractors existing in the minifundio districts of the Central and Northern Coast increased by 15,101 units.
17. There has been a progressive substitution of native cows with Holstein, leaving the former increasingly restricted to marginal areas.
18. This was done by either tractors or motocultivators or by resorting to hired machines, which was a very common practice. However, it is worth noting that in certain areas many dairy cows continued to be used in field work, although such work was lighter than before.
19. It is worth noting that until the 1960's the minifundio areas were the largest suppliers of milk in the dairy industry, which usually undervalued milk and paid with great delays.

References

Baptista, F. Oliveira, *Politica Agraria* (anos trinta -1974), Lisboa, 1984.

Baptista, F. Oliveira, "Estrutura da producao agricola Portugal-um panorama do ultimo meio seculo" in *Pensamiento Iberoamericano* n-o 8 Julio-Deciembre 1985, pp. 331-365.

Caldas, J. Castro, "Caseiros do Alto Minho-Adaptacao e declinio" in *Revista Critica de Ciencias Sociais,* n-o. 7/8 Dezembro 1981, pp. 203-216.

Cordovil, F. Cabral, "Transformacoes da estrutura das exploracoes agricolas em Portugal nas ultimas tres decadas (1950-1980) e efeitos previsiveis da adesao a CEE" in *Economia e Socialismo,* n-o. 61 Abril-Junho 1984.

Cordovil, F.C., Antunes, C., Ccmpos, J., Santandre, J, *Contas de rendimento das familias no Continente, Distritos, Sub regioes e Regioes,* Lisboa, IACEP-GEBEL, n.d.

Estacio, F., Alves, A.M., Avillez, F., "Evolucao da agricultura portuguesa has ultimas decadas. Uma perspectiva economica", Comunicacao apresentada no Seminario A Agricultura Portuguesa, ISA, Lisboa 26-30 Novembro 1984.

Lacticoop (ed.), *Intersificacao da producao leiteira* (colectanea de taxtos), Aveiro, 1983.

Moreira. M. Belo, A pequena producao e as salas colectivas de ordenha mecanica. Elementos de caracterizacao do litoral e do Interior do distrito de Aveiro, Lisboa, Centro de Economia Agraria e Sociologia Rural, INIC, Novembro 1982.

Moreira, M. Belo, *L' economie et la politique laitiere au Portugal (1926-1981),* These de doctorat de 3. eme cycle, IREP, Grenoble, 1984.

Moreira, M. Belo, "An outline on portuguese agriculture" in Facts and Figures About Rural Portugal, SPER-13th European Congress for Rural Sociology, Braga, 1986.

Rodrigues, Manuel, "Problemas da producao de leite no Noroeste" in a *AGROS* no. 3, Ano XXXIV, Maio-Junho 1951, pp. 140-158.

Rolo, J.C., Cardoso, F.M.V., Cordovil, F.C., Lopes, R.M.G. *Produto Agricola Bruto no Continente, Regioes e Concelhos (mec a do trienio 1979-1980-1981),* SIP-DCP, Dezembro 1985.

8

THE CAPITALIST IMPLANTATION BY SUBSTITUTION OF PRODUCTIVE FORMS IN VENEZUELAN AGRICULTURE[1]

JOSEFINA RIOS DE HERNANDEZ

Since the beginning of the 1920's, Venezuelan society experienced a transformation process grounded in the development of the oil industry and in the State performance as a distributor of the increasing fiscal revenue. In agriculture a new agrarian structure emerged based on the predominance of capitalist relations of production organized in a capitalist productive form.[2] Contrary to the previous historical model of export agriculture [3], the new one was dependent upon the internal market which grew as a result of overall changes in the Venezuelan economy. The capitalist productive form developed initially in lands that had not been previously occupied and it was not until the 1960's that the changes extended to all historical agriculture.

It should be emphasized that capitalist agriculture in Venezuela was not a result of the internal evolution of the existing productive relations, but it developed as a consequence of external conditions. State action and agroindustrial activity were the most important factors in that sense.[4] Due to this fact we use the word *implantation* to identify the particular process of capitalist agriculture development in Venezuela.

Capitalist implantation was a complex process that took different routes. Colonization was the most important among them due to its earlier occurrence and its widespread geographical coverage. It contributed to a shift in of the main centres of agricultural activity from the coast-mountainous zone to the western high plains and to the basin of the Maracaibo lake as well. It should be noted that extension of agricultural frontiers was associated with an important population migration.[5]

Implantation also occurred through substitution of precapitalist productive forms with capitalist ones. With less geographical scope and influence in the agricultural development, that process took place in lands already occupied by *hacienda hato* or mercantile-family production.[6] In all cases, substitution meant changes in technical conditions, in labour and exchange relations, and in patterns of land tenancy, as well.

This paper refers to the substitution process considering the most relevant cases. We classified them according to the importance of the role played both by state agencies and by private entrepreneurs. Even though other criteria may be used to distinguish the particularities of that process as it evolved in the historical areas, the one indicated allows us to emphasize a characteristic feature of capitalist agricultural development in Venezuela, that is, the particular ways in which private and public initiative came together to push agricultural development.

The Substitution Process by Private Entrepreneurs

Implantation of capitalist agriculture through the substitution of previous productive forms, in this case, was promoted and carried on by private agents although they might have had State support. It took place as a result of the introduction of a mechanized technological pattern and, in some agricultural areas (mainly in northern-centre and western regions of the country), also as a consequence of partial changes in land tenancy.

The substitution process began in the late 1930's in the basin of the Valencia Lake. In the forties, it expanded toward San Carlos, in Cojedes state, and into some areas of the Yaracuy state. In the following decade, substitution became stronger as a consequence of the construction of large watering systems in the high plains of Guarico and Portuguesa states and of the development of tobacco agro-industry. Here we will consider the cases of Valencia Lake and tobacco substitution.

In the first case, *haciendas* underwent a transformation of previous cultivation activities and of the productive process itself. This change was a result of the growing demand for agricultural goods that accompanied an intense urbanization process in the northern-central region, along with the development of an increasing purchasing power and new consumption patterns within the population located in this region.

Those haciendas had comparative advantages with regard to types of soil, proximity to the markets of the country, and availability of lands ready for cultivation under the new technological pattern. Additionally, they had relatively good road and transportation systems; the first two dams of the country were built nearby; and electric service was improved, contributing to an increase in the irrigation potential of the area.

With new crops like maize, cotton, sugarcane, tobacco, tomato etc., and livestock exploitation for milk production, those haciendas changed the monoproduction and export orientation that historically had defined its character as a productive form. A diversified production was now mainly destined to the internal market.

The new crops had better commercial advantages. For example, maize and cotton had a secure market, as well a minimum sale price. By Governmental disposition the textile industries were forced to buy the national production at fixed prices, before having resort to foreign cotton. In the case of maize, the "Banco Agricola y Pecuario", a governmental agency, offered to buy all crop transported to some storage places at a fixed minimum price. Besides, industries of the northern-central region bought maize for preparing animal feed. The localization of storage places and processing industries in this region gave additional advantages concerning the transportation of the crops.

The development of new products gained a renewed momentum with governmental policies of the mid-forties encouraging the development of capitalist agriculture, through particular plans for different products. The most relevant were cotton and sugarcane plans. The latter included the building of sugar mills, as well as the establishment of national organization for dealing with sugar commercialization.

The expansion of sugarcane production since the beginning of the fifties was mainly a result of a substitution process. In northern-central region it was stimulated by a sugar mill "El Palmar", that was built with private capital. It offered sugarcane cultivators a secure market by fixing purchase quota, and provided them with financial and technical assistance. Similarly, livestock exploitation received a special stimulus when pasteurization plants were established in that region.

The use of a new technological pattern contributed to modifying in a significant way, the productive process. Mechanical procedures, chemical inputs, and intensive land exploitation practices were the main features of the pattern, which became generalized for the new crops. It introduced technological innovations like new varieties of crops, chemical fertilizer, improved seeds and insecticide plague control, new and healthy breeds of cattle, pasture cultivation, processed foods, and control of animal reproduction, among others.

The main actors of these changes were *hacendados* with economic capacity to invest on their own in the new activities. Frequently the hacendado associated with other agents with skill and experience in the new technological pattern. Most of them were migrants from the Canarian Islands in Spain, who often took charge of the productive process and participated as labour force, sharing with the hacendado a fixed percentage on the total benefits.

The labour force came, basically, from the population that had been previously bound to the *hacienda*. Some haciendas kept until the end of the fifties the traditional "peonaje" labour relationships on the basis that the worker held a plot of land in the hacienda without paying for it, and at the same time he obtained a remuneration when he participated in the productive process commanded by the *hacendado*.

However, the new activities were mainly carried out by day labourers (jornaleros) and full-time wage-earners. In the first instance, which was also typical of the hacienda, the worker had some means of production independent of the hacendado and worked temporarily on the hacienda in order to obtain additional income. In the other case, full-time workers were individuals whose reproduction depended primarily on the income earned on the hacienda, as alternatives were lacking or did not generate sufficient income.

Most day labourers and wage-earners lived in areas near the hacienda. However, in times of higher demands of labour force, such as the harvest time, it was necessary to have migrant workers from other zones of the country. Foreign migrants, mainly from the Canarian Islands, also participated as permanent labour force. Because of their skill and experience, these workers were often engaged to supervise different stages of the productive process.

Capitalist agriculture based on mild tobacco exploitation developed basically in the fifties with the introduction of "virginia" and "burley" varieties and new technological procedures for cultivating and processing the product.

Two agro-industries gave impetus to tobacco exploitation in those years. A "Tabacalera Bigott" and "Compania Anonima Tabacalera Nacional (CATANA). "In the next decade it was encouraged by protectionist and import substitution policies. High taxes and licence for import of tobacco became the main steps in that direction. Besides the industries were obliged to account for 50 percent of the financial funds for cultivation. As a result, in 1966 mild tobacco production covered the growing national consumption.

The irrigation requirements of tobacco conditioned its localization in plain low lands with potential irrigation, such as the basin of the Valencia Lake and the fertile plains of rivers in the central region (Tucunemo, Orituco, Guarico, and Cojedes). As well, tobacco exploitation expanded to the plains of the Acarigua and Guanare rivers in the western region.

Plain lands of haciendas in the basin of Valencia Lake were dedicated to tobacco exploitation that, in the forties, was more profitable than export products, such as as coffee and cocoa. In Guarico state, exploitation of tobacco began in the early fifties in lands that had been occupied by small exploitation of black tobacco, associated with other crops in mercantile family production units; but most of the land had an extensive cattle raising activity organized in *hatos*. In the same way, in the low lands of the Acarigua river (La Mision), there already was black tobacco produced by traditional means, as well as in mercantile-family production units.

All the above cases had one factor in common, that new investment and technological pattern had changed the existent production conditions. Although many haciendas and hatos owners used part of their lands for tobacco cultivation, new producers emerged that purchased or rented the land. In the same way, former land owners, some from the previous mercantile-family production, could extend their properties incorporating lands adequate for tobacco exploitation. In spite of this, most of tobacco production was organized in small units of production (no more than 100 hectares), comparing the extension of haciendas and hatos, and was frequently associated with another crop.

The tobacco industry not only promoted the extension of cultivated lands, but also stablished a close relation with agricultural producers through supervised financial support and purchase quotas. In the same way, industry determined the technological pattern to be used in cultivation, fixed criteria for classifying tobacco leaf, and established purchase prices for the producers.

In order to open credit, the tobacco industry did not required the producer to have previous experience with tobacco, since it involved training and continuous technical assistance at each step of the productive process. Also the industry gave financial support to buy land and to build structures for processing tobacco.

The commercialization of tobacco was realized by the industry by means of previously fixed quotas that determined the variety to be cultivated and the purchase price, depending on quality. Tobacco producers were assigned the industries that exercised a direct control, taking care that each producer was working only with one of them. This reflected the oligopolistic character of the tobacco market and the strongly subordinate link the producers had with the agro-industry.

There were different types of producers according to the availability of resources and the relations they established with the industry. On the one hand were those producers with land, a quota from the industry, and a supervised credit, carrying on production under the industry's direct control. On the other were those producers, working in association with others, under different forms of sharing costs, risks, and benefits that were established at the beginning of the productive process, according to the availability of resources of each participant.

The association of producers had been organized since the very beginning of the substitution process, and became generalized as the tobacco demand increased. This form of organizing the exploitation of tobacco was considered essential because of the difficulties in obtaining quotas from the industry, the high amount of investment required, and the complexity of the productive process.

Given the characteristics of the technological pattern, cultivation and processing of tobacco were laborious activities that demand a considerable number of workers for the different tasks (like sowing the plants, cultivating them, keeping up the crop, drying and baking the tobacco leaves, etc.). The high requirements of labour force made tobacco production an important source of employment in the zones where a relatively abundant population already existed

In the context of the technical conditions already indicated, organization of the productive process necessitated the establishment of new labour relationships that modified the existent ones: *Peonaje* was replaced by day labourer and wage-earner work relationships.

Substitution Process by State Intervention

This process involved the development of private capitalist agriculture promoted and supported by State intervention as part of agriculture modernization. In this sense, the State outlined development projects to be carried out in lands already occupied by alternative productive forms that were changed and substituted with new ones.

The first sign of substitution by State intervention was the organization of agricultural colonies in the thirties, using lands acquired by the State from private owners. These lands were assigned to new producers, including the foreign migrant population. That was the case of "Chirgua" and "Mendoza" colonies in Carabobo and Aragua states, both in the northern-central region.

Here must be mentioned an experiment carried on by a State agency during the 1945-48 period, through an anonymous company called "Agropecuaria Maracay". This company managed a group of haciendas localized in the Aragua state, which had belonged to the former dictator of Venezuela, Juan Vicente Gomez, and were confiscated by the Government at his death. The haciendas were transformed into model productive units for livestock intensive exploitation. However, the experiment was given up when the Government changed in 1948.

The main cases of substitution by State intervention occurred in the 1950 decade, linked to rice exploitation. The Guarico Water System and the agrarian development in Acarigua-Araure zone were the most important projects in the country in those years, as well as being pioneers in this type of agriculture organization. Both took place within the Rice National Plan carried out by the Government.

The first project was part of the Water System started in 1948, and was designed to be concluded in 1970. In this full care was taken of population settlement pattern and production and marketing programmes, with the Government responsible for bringing the financial and technological support needed to the productive process. At the same time, it was to provide health, education, and recreation services.

The Guarico Water System used the Tiznados and Guarico rivers to irrigate the vast agricultural plain land and to control natural floods in the rivers. The water works were started in 1954 and completed three years later. The dam covered 23,000 ha with an irrigation area of 110,000 ha. Initially 70 percent of the land was going to be used for livestock rearing and the rest for agriculture. However, more than 75 percent of the land had clay soil with drainage difficulties, so it was used for rice plantation, the 25 percent of the land left being used for fruits, sesame seeds, and forage.

With the Water System and new techniques of cultivation, the zone became virtually a rice bowl. In a decade, Guarico and Portuguesa states constituted the main rice producers of the country. In 1960 the two account for 50 percent of the national crop, and 90 percent of Guarico state production came from the water system that had 85 percent of the cultivated land.

For building the Water System, the Government expropriated from private owners nearly 200,000 ha. Most of the land was occupied by large hatos, and the rest was communal property land held by mercantile family producers.

Initially, the project had an entrepreneurial character. In this sense, between 1957 and 1959 nearly 20,000 ha were distributed in plots that varied from 150 to 270 ha. Most of them were assigned according to technical criteria, in order to select producers that had previously been engaged in some type of agricultural modernization. Most of the hatos owners did not participate in the project, although some of their descendants, with agricultural technical skill, did later. Besides, political influences also were taken into account for selecting some of the producers. By those criteria, previous family mercantile producers were not eligible to manage a plot. As they were dispossessed of the land, most of them had no alternative but to become the labour force of the new producers.

After 1959 the population settlement pattern began to change with the organization of the so-called "peasants entrepreneurs", integrated by previous family producers of the zone. Each family obtained 10 to 20 ha to be exploited not individually, but as part of a larger productive unit with 150 ha and 10 families as a minimum.

Though producers began to establish themselves in 1957, it was a decade later when the Government actually sold them the land and gave property titles to some of them, as a result of the claims made by the Water System Association (APROSIGUA). In this arrangement the amount to be paid was reduced according to the family size (5 percent for each child and 15 percent for wife).

The building of the water system and the development of capitalist agriculture attracted migrant population that changed the demography of the zone. The absolute intercensus growth between 1950 and 1961 period was 16,948 inhabitants, which meant a relative population increase of 57.8 percent.

The strong financial and technical support given by the State helped the producers to organize a productive activity using mechanical implements and chemical inputs. Although family labour relationship had been a condition for obtaining a plot, the technical means used increasingly caused the replacement of the family labour relationship by day-labourers and wage-earners. This change affected not only the larger producers but the small ones organized in "peasants entrepreneurs".

The Government also carried on a marketing arrangement politic aimed at securing benefits, and increasing production and productivity to the producer. In the same way the high quality of rice production led the Government since 1960 to buy all the harvest at higher prices than the fixed minimum.

In those conditions in Guarico Water System was constituted a group of medium agricultural entrepreneurs (also family producers) strongly linked to governmental support. They became a sort of government agent, responsible for extending capitalist agriculture, since the Government brought them all the resources required for agricultural production and created facilities for its continuity. In spite of the fact that they were private producers, their dependency on the government did not decrease in a significant way.

On the other hand, the settlement of capitalist agriculture by substitution of productive forms, with State intervention in Acarigua-Araure zone had a different character. In this case an entrepreneur group that emerged in a short time obtained an important economic power and could work with more independence, though they continued receiving government support.

The agricultural development in the zone started at the end of the forties, as an integral project that put together different governmental policies. The government intended to put forward a set of financial, technical, and marketing actions that might enhance private initiative to establish productive units under new technological and management patterns.

The importance that the zone acquired was reflected not only in the rice production level, but also in the fact that the entrepreneurs group could rapidly consolidate and extend its economic power in agriculture and, at the same time, participate in other related economic activities. Acarigua city became the centre of that development with a dynamism not known before.

The government's Rice Plan with supervised credits was the base of the substitution process in Acarigua-Araure. It was carried out from 1947 by CVF (Venezuelan Corporation of Foment), stimulated by the fact that national rice production could not cover, at that time, national consumption (between 15 and 25 percent of apparent consumption), while prices in the international market were very high. The Acarigua-Araure zone had advantages for rice exploitation—adequate land and good necessary conditions, compared with similar lands still affected by endemic illness.

Initially, technical skill in mechanical agriculture was the only condition taken into account for selecting producers to be given the credits. In this sense, three engineers and 14 agricultural technicians, who were working in different governmental agriculture agencies, were the first to obtain credits for rice exploitation in that zone. Among them only one had already land in the zone with extensive cattle raising, and three lived in the Acarigua-Araure zone. Most of the people selected came from the northern-central region, and had no land at all. So, the settlement of capitalist agriculture under the Rice Plan meant a change in land tenancy and the participation of new social agents as producers.

Having obtained the credit, the potential producer had to look for land on his own. It was not an easy task because most of the land for rice exploitation, although municipal property, was occupied by private agents, members of the Municipal Council. when they refused to rent our part of that land, some of the "technicians" managed to participate in that council, and from this position they expropriated the former occupants and rented out the lands to the new producers; each one was given 200 ha for a period of five years, to be renewed by equal periods when the producer paid the rent on time. Later on, new producers were given the benefit of the Rice Plan, but the same technical criterion were not made the basis of allocations.

In eight years there were no more plains lands for rice exploitation, so nearby areas that previously had only lumbering exploitation were used. The use of these lands for agriculture was more difficult than the former, because good investment for getting the land ready for the use of mechanical means was necessary. For this reason it took more time to develop agriculture in this area, though it had better advantages due to the good quality of the soil.

The Municipal Council put no limits to the extension of land to be rented in that area, so very soon some of the producers, including the former (technicians) organized large units, some of them with more than 1,000 ha.

Initially, rice exploitation was developed with family work, combined with day-labourers and wage-earners. The producer not only organized the productive process, but he worked directly on the different tasks. Population settled in the zone or coming from areas nearby, in Falcon and Lara States, were the source of labour, which was secured by paying higher remunerations than what other producers in the zone paid.

The first group of producers started rice exploitation with a credit at low rates (5,6 percent), to be paid to the CVF directly with the harvest. First, they received Rs. 200,000. Part of it was meant for the maintenance of the producer and his family until the first crop was ready. The rest of the credit was for machinery and instruments for agriculture exploitation. As the producer returned the credit as agreed to he could get a higher one for extending cultivated lands.

At the same time, this group had other advantages that contributed to the establishment, in a short time, of a capitalist form of production. Along with the Rice Plan the government started a programme for improving roads, bridges, and other structures and establishing services to support agricultural development. A manifold technical assistance coming from different sources, public and private, helped to consolidate the know-how about the new technological pattern among the producers. Also, they managed to organize an association that articulated their economic and social interests, mainly oriented to pressure the government to provide an adequate and ready support.

On the basis of this development, in the Acarigua-Araure zone was constituted an entrepreneur group that succeeded in expanding its activities to add other products and activities. They introduced sesame seeds with new varieties that increased the levels of production. It was cultivated in the same land, using the free time left from rice production. Later on, when rice exploitation started to have difficulties, they also added cotton, maize, and sugarcane. In the same way, some of them developed livestock rearing with a double purpose: milk and meat.

The entrepreneur group diversified its economic interests, expanding agricultural production and participating in agro-industry and financial activities. In that way, they also gained economic support for the agricultural activities and, in many cases, were able to extend their influence to other areas.

The diversification of agricultural products brought these producers in touch with different financial sources, not only governmental agencies, but also private institutions. At the same time, by using multiple public and private channels of marketing, they also became less dependent on the State.

As the technological patterns that were basic to this type of development became increasingly complex, agriculture and livestock exploitation became labour saving activities. The modifications of technical conditions contributed to an increasing and continuous capital accumulation that led the producers to replace the familiar labour force that they had at the start. They retained the supervisory tasks while the routine production process was mainly transferred to day labourers and wage-earners.

To summarize, the particular experience of Acarigua-Araure entrepreneurs represents an example of substitution by government intervention that rapidly consolidated an economic power which gradually became an autonomous group. The combination of favourable conditions added to personal characteristics of the producers explain the particular development of capitalist agriculture in this zone.

Notes

1. This paper is a summary taken from the work: *The process of constitution of the Capitalist Productive Form*, which is part of the findings of the research: "The Relations of Production in the Venezuelan Agriculture. 1900-1980". These are based on oral testimonies from key informants specially chosen from among the protagonists of the facts that are here described. The testimonies were obtained through open and nonstructured interviews in different regions of social agents that occupied distinct positions in the agriculture production. Given the limitations of this paper the testimonies were not included but they are extensively used in the original texts.
2. The theoretical and methodological foundations for concepts and categories used in this research are in a document specially addressed to that purpose. Nelson Prato Barboza: "Conceptos, Categorias y Criterios Teoricos Metodologicos de la Investigacion". First Part, Chapter 1. Area Sociohistorica de CENDES. Caracas, 1985.
3. The term refers to agriculture in the agroexportation period, developed from the 16th century up the first decades of the present century.
4. The role played by the agroindustry as central axis of the agriculture dynamic is analyzed in Josefina de Hernandez y Nelson Prato, "De la Agroexportacion to the agroindustry". Area Sociohistorica CENDES. Caracas, 1985.
5. See: *The Capitalist Implantation in Venezuela Agriculture*. Nelson Prato Barboza. Area Sociohistorica del CENDES. Paper presented at the XI World Congress of Sociology. New Delhi, India, August, 1986.
6. These productive forms are extensively considered in the following works: Josefina de Hernandez, *La Hacienda Venezolana. Una Vision a traves del testimonio oral.* (in press). Gaston Carvallo, *El Hato Venezolano.* (1900-1980). Editorial TROPYKOS. Caracas, 1986. Josefina de hernandez y Nelson Prato, "La Forma Productiva Familiar Mercantil", Area Sociohistorica, CENDES-UCV Caracas, 1985.

9

THE CAPITALIST IMPLANTATION BY COLONIZATION IN VENEZUELAN AGRICULTURE[1]

NELSON PRATO BARBOSA

From the decade of the 20's there is a process of transformation in Venezuela, affecting agriculture and basically driven out by the petroleum exploitation. This process has accelerated since the 40's with the changes in income distribution and growing urbanization that go along with the new petroleum substract of the national economy and the changes in the population consumption patterns. In the 60's these changes become faster, generating a new agrarian structure commanded by the State policies and the predominance of capitalist relations of production structured in the capitalistic productive form.[2]

The functioning of productive forms is based on a mode of appropriation of the surplus-product in which the worker is separated from the means of production. Consequently, his own reproduction and the one concerning to his family depends on a wage regime.[3]

The mechanical technology, bioenergetical and chemical inputs, and intensive land exploitation practices required for the capitalist productive form were advancing, step by step, depending on the kind of crops, availability of capital, and the historical background of the area where the process was carried out.

The capitalist production form depends, essentially, on the wage-earner, along with the family and the day labourer (jornaleros). This form of labour was characteristic of previous productive forms, such as mercantile-family productive form and hacienda productive form.[4] Their importance changed through the history and the different scopes where they existed. However, the historical tendency is to strengthen the salaried labour system.

In spite of the preceding agricultural structure, mainly based on the exploitation of coffee and cocoa, capitalist agriculture was advocated to the internal market which was growing as a consequence of the petroleum impact on the country's economic structure. In this context, the capitalist entrepreneurs became involved in a system of exchange with the agro-commercial and agro-industrial agents of circulation. The agro-industry has become the articulator axis of the relation of production in the Venezuelan agricultural and would play the central role in explaining the new agricultural dynamics of the country.[5]

The conformation of the capitalist productive form in Venezuela can be seen as a consequence of different and confluent historical processes in which the State plays a very important role. The implantation process referred to above means that the new dynamics is not a product of internal characteristics of historical agriculture but the result of external influences. In this process the State became the fundamental agent not only in the generation of conditions for production and reproduction, as it is generally conceived, but as an active actor.

There are two ways in which capitalist relations of production are implanted in Venezuela; first, by the appropriation of free lands expanding the agricultural frontiers; second, changing the previous productive form to the new kinds of relationships.

The private initiative as well as the State intervention were the modes that contributed to the conformation of this process. This paper will focus on colonization, given the great importance that it had in the capitalist productive form in the Venezuelan agriculture.

Implantation by Colonization

Colonization, as a process of agricultural frontiers expansion associated with gross internal population movements, was the main core of the earlier capitalist relation of production in Venezuela. It was followed by the setting up of the mercantile-family units of production. This particular articulated combination of relations of production stimulated capitalist development, not only because of the different agroecological conditions, but also because of the advantages of the automatized technological pattern and the State supportive interventionism.

The capitalist implantation by colonization occurred mainly in the Lake of the Maracaibo Basin, in the State of Zulia, and in the high west Plains, States of Portuguesa and Barinas.

In spite of the fact that it lasted for 40 years, the beginning of the colonization process, in each of the above mentioned regions occurred at different times and exhibited regional specificities. These specificities depended on the conditions in which the occupation of the territory took palce, as we are about to see.

The capitalist colonization process began during the 20's, in the northeast of the Lake of Maraçaibo Basin, mainly in the Distrito Perija; during the next 10 years it expanded to the south, mainly to the Distrito Colon, and this pattern was reinforced in the 50's, when the Distrito Sucre, in the southeast, was incorporated. Later on, during the 60's and 70's, the implantation process expanded to the east.

In spite of the land occupation and the lumbering, it was not until the 40's when the process of colonization became more significant in the high west Plains.

Seen as a whole, the process of capitalist implantation by colonization, advanced slowly up to the end of the 40's when it began to intensify and expand its area of influence and its agricultural and geographical scope. These differences in the process, rhythm, and expansion are mainly related to the kind of social actors that took part in the process and their performance. With the available financial resources, those actors actually participated in land conditioning in order to establish the production units.

In areas under colonization, two modes that influenced the final capitalist implanted production can be identified. Fundamentally, it was initiated and strengthened by private groups on their own in the Lake of Maracaibo Basin; at the high west Plains, the State intervention seemed decisive for the advance of the process.

As we shall see, in the first mode the population settlement process and the consolidation of production units generated a *primary process of accumulation;*[6] in the second mode the population was settled on land previously conditioned by the State with available technological and financial resources, within the framework of a policy oriented toward capitalist development in agriculture. What follows will be addressed to review each one of these capitalist colonization modes.

Colonization by Private Initiative

The case of the Lake of Maracaibo Basin, mentioned above, is the most significant in this mode of colonization, and it was based on livestock exploitation with a double purpose: milk and meat.

The analysis of this case is important, first, because it began within the framework of an agroexportation economy. By that time, the new implanted units were oriented to meet regional demands that were enhanced for the exporter role that the Lake of Maracaibo performed in the Andean coffee commercialization (The Andes was the principal coffee producer in Venezuela). In the second place, this case can be differentiated from others because it had no ties with the functioning of the productive forms of the historical agriculture, in contrast to what happened with the rest of the agricultural processes that seriously declined. This case became one of the successful agricultural processes for implantation of the new capitalist relations of production. In the thrid place, this process of capitalist colonialism occurs in the same geographical area where the most important early petroleum exploitation was just taking place. Furthermore, it became the first case of agricultural expansion under the new social conditions of the Venezuelan society.

The base of the colonization process during its initiation period is the settlement of colonists and appropriation of the new lands. The beginning of the appropriation process occurred under a relative lack of resources, but with no limits to possession of the land except for bordering and deforesting.

Although most of the occupied lands were freelands, "ejidales" (municipal lands), and national lands, the petroleum companies took some of them from indian groups in order to achieve their petroleum exploration activities. In this sense, the process of colonialism was not free from conflicts among the colonists and the indian groups, due to the bordering and the sizing of units of production.

Within the context of an unlimited land supply, the pioneers' capacity for appropriation, expansion, and accumulation depended on the availability of resources. This fact contributed to what we call a process of primary accumulation, which developed in a different manner than in the other areas. The creation of varied relations of labour, such as family work, day labourer work and in some cases forced labour, combined with wage-earner work, were also some of the mechanisms to accomplish the process of primary accumulation. As long as it was socially differentiated, it facilitated a concentrated pattern of land property among the colonists. It also contributed to spreading and consolidating capitalist development against mercantile-family production units. Buying and selling, the dislodgment of colonists, and free expansion of the first properties were the mechanisms through which the concentration process occurred. Although work relationships within the family were being replaced, they retained some ties with the capitalist implantation as in the case of supervision tasks (especially those that involved father-son relationships).

In the beginning, indian population was compelled into forced labour, but this was abolished in the 30's with the death of the dictator Juan Vicente Gomez. We might remark that a significant part of the indian population offered strong resistance to the new slave-like conditions of work. An analysis of forced labour shows that the scarcity of manual labour was characteristic of colonization in the region of Zulia, creating differential requirements for manual labour and furthermore, enhancing the exploitation of livestock instead of agriculture. Nevertheless, it is important to point out that when the primary accumulation process began, it was supported by the commercialization of agricultural products. These areas of production later on generated the creation of stockyards.

The capitalist units of production that came up through these processes were based on the work of day labourers and wage-earners, where the latter was favoured as a consequence of transformation in the political conditions, the development of a Venezuelan capitalist society, and the expansion of the capitalist units of production.

That particular combination did also strengthen the wage-earner relationships and the growing importance of the immigrant workers (coming from Los Andes, the State of Falcon, the north part of Zulia and from Colombia), who were directly engaged in the production process and also participated in the foundation of the new units.

There were some other factors which contributed to reinforce the private form of capitalist colonization in the Venezuelan agriculture: the commercial development in the State of Zulia made possible the expansion and multiplication of the livestock settlements. The same role was played by railroads, lake and river transportation, and the local market for milk and its products. Automobile transportation was developed later (during the 50's). In this decade the opening of the Panamerican Road highly contributed to the consolidation of the process.

The developing of the agroindustry of milk is equally relevant. The setting up of the milk processing plants in the 20's clearly contributed to the specialization of the livestock production initially started with the cattlemen's capital. The State intervention occurred during the 40's by stimulating the foreign capital investment, and by the installation of

"NESTLE"). In the next decade, this process became more important with the installation of multi-financed milk processing plants (private local, public, and foreign investments).

In the same sense, the State credit policies during the 40's were aimed at technological renewal of the production units, introduction of new breeds of livestock, and control of prices. Also, public appeals for consumption of processed milk played a significant role.

The expansion of the agricultural industry in the region of Zulia (later on it spread throughout the country) contributed to facilitating financial support for the producers and also provided better technical advice, making cattlemen's work more productive.

The State intervensionism contributed to enlarging the milk market, and concentrating and integrating the capital, as can be seen in the case of INDULAC, the firm that finally became a State-private financed enterprise.

Also, the State policies stimulated meat production without any price control, making it more profitable. This led to a double use of livestock production in an attempt to balance the benefits in the functioning of internal market.

The consequences can be summarized as follows:

1. Intensive capital investment and intensive use of lands;
2. Incorporation of new patterns of technology;[8]
3. Political and economic strengthening of local cattlemen's groups which became very important at the national level.

Colonization with State Intervention

The most significant case in this mode of colonization was the Turen Agricultural Project (in the State of Portuguesa), originally designed in 1949 and started in 1950. Today, Turen constitutes the most important State-private financed project in bordering agricultural lands; this region is commonly known as the "Granary of Venezuela".[9]

Most of the occupied lands were municipal property (ejidos), and since the 20's certain areas were under a special regime of permission for lumbering. But, it was not until 1947 that the lumbering industry started to generate the first spontaneous internal migration. These people started mercantile-family units and some of them were municipal tax payers.

In 1938, the State stablished the "Instituto Tecnico de Inmigration y

In 1938, the State stablished the "Instituto Tecnico de Inmigration y Colonizacion" (ITIC) to begin planned colonization projects. In 1949, the ITIC became the "Instituto Agrario Nacional" (IAN). By that time, there were 19 colonies in the country, all of them based on the substitution of previous productive forms.[10]

The Turen Project had special advising from consultants from the United States, and it was planned to change about 27,000 ha of jungle into productive lands where 6,000 families would be settled, 20 percent of them from foreign countries, mainly from Europe.

The main objective of this project was to produce agricultural inputs for the incipient industrial activities by introducing and diffusing new kinds of products and technology. The units of production were initially created with enough space for the family-labour form, in order to produce in a riskless and mechanized way.

The plots to be used were deforested; however, 20,000 hectares were reserved to guarantee the ecological balance in the area. Moreover, fully equipped housing and necessary services were provided, as well as food, until the colonists obtained their first crop. They also received 27,000 Bs. (U.S. $ 9,000 approx.) in machinery and instruments for agriculture, provided through the State credit plans under long-term conditions of payment.

The State also provided medical and educational services, church, roads, and an airport, and guaranteed commercialization of crops by building granaries and warehouses.

The project's progress required the dislodging of the inhabitants of the zone, who were later to be employed as workers. These rehabitants had fiercely resisted the first attempts to enlist their cooperation.

By 1950 about 450 houses were built, and the project directed its attention to foreign groups: Germans, Italians, and Spaniards among others, who were given plots of about 50,000 ha. Some Venezuelans from other regions were also benefited, but with smaller parcels of land. In 1954 the biggest plots were occupied by foreigners in about 86 percent;[11] meanwhile, 90 percent of the natives were in small parcels of land. This situation generated social conflicts which required repressive intervention by the State.

From the very beginning the IAN controlled the units, step by step, from the kind of crops they were raising up to the commercialization of the crops.

From 1954 the process of colonization was extended to all the areas of Turen, but without having State intervention at the same level as in the case of the "Agricultural Colony of Turen". Actually, the new colonists started to organize themselves in order to discuss the benefits they would obtain from the State.

The process always did require abundant labour force, and this demand made the offer varied in terms of labour relationships: family work, day-labourer work, and wage-earner work.

The families benefited with plots were particularly engaged in the mechanized work, while the manual work was done by day labourers and wage-earning workers. The technical requirements of the production process imposed the need of a specialized work force and opened a work market for immigrants who began to substitute the role of the family work. Also as the production process was specialized, the demand for qualified workers grew.

From its very beginning, Turen was aimed at satisfying the needs of industrial processing, especially those crops that required short-time cycles. This process of specialization determined the agrarian concentration of production by zones and regions, as can be seen in the case of Altos Llanos Occidentales, concentrated on producing rice, cotton, corn, and sesame seeds

The State also created the mechanisms of commercialization such as Banco Agricola y Pecuario, CORPOMERCADEO, BANDAGRO, etc. The next step was a process of capital concentration as the successful colonists began to buy the land from those who failed or could not maintain the necessary rate of benefits, making their own land plots to expand up to 250 ha. At the same time, the successful colonists made an effort to free themselves from State interventionism but did not succeed entirely as the State was directly involved in most of the financial support.

The Agrarian Reform policy also supported the small plot owners during the 60's. However, it must be taken into account that the main purpose of the Agrarian Reform was to organize the land workers under the influence of the largest political parties (Accion Democratica and COPEI), dislodging some colonists and relocating some others.

From the 60's onwards the Capitalist implantation by colonization was barely consolidated and this permitted it to extend to the lands that were still free. During the 70's the CODESUR Programme was specially aimed at incorporating the southern jungles (Guayana, Delta Amacuro and Amazonas) in the process. However, this programme was soon given up because it did not have the expected success and today, the process is still going on in a more unstructured form, though the State intervention retains its importance to ensure the process.

Notes

1. This paper is based on the chapter named: "El Proceso the Constitucion de la Forma Productiva Capitalista", which is part of the project: "Las Relaciones de Produccion en la Agricultura Venezolana 1900-1980", under the coordination of Prof. Josefina de Hernandez. Ocarina Castillo and Gaston Carvallo also participated in the research as members of the Sociohistorical Unit at CENDES. It was partially financed by CONICIT and has obtained the annual award to social research from the Professors' Association of the Universidad Central de Venezuela in 1985. The findings presented are based on oral testimonies of key informants specially chosen among the protagonists of the facts that are here described. Given the limitations of this paper these testimonies could not be included. The oral testimonies were obtained through open and nonstructured interviews in different regions taking into account the role performed by each of the social agents in the productive process.
2. The theoretical and methodological foundations for concepts and categories used in this research are in a chapter specially addressed to that purpose. See: Nelson Prato Barbosa "Conceptos, Categorias y Criterios Teorico metodologicos de la Investigacion". First part, chapter 1. Area Sociohistórica de CENDES. Caracas, 1985.
3. The concept of wage or salary is a specific kind of remuneration. It must be differentiated from others that could also include monetary owner-worker relationships. Prato, *op. cit.*
4. These concepts are extensively developed in different papers on the hacienda productive form and the mercantile-familiar productive form. Cfr.: Josefina de Hernandez y Nelson Prato: "La Forma Productiva Familiar Mercantil", Area Sociohistorica, Caracas, 1985. Josefina de Hernandez: "La Forma Productiva Hacienda", Area Sociohistorica. Caracas, 1985.
5. The changes in the articulator axis of the relation of production in Venezuelan agriculture are developed in: Josefina de Hernandez y Nelson Prato: "De la Agroexportacion a la Agroindustria", Area Sociohistorica. CENDES. Caracas, 1985.
6. The concept of primary accumulation is here used in a different sense to that of original accumulation which specifically refers to the constitution of European capitalism. It has been specially constructed to describe the conditions of capital accumulation in the Venezuelan Agriculture. Prato, *op.cit.*
7. The concept of forced labour is specifically referred to the process of colonization by private initiative in that period. See: Prato, *op. cit.*
8. Within the technological package it is necessary to add: pasture harvest, breed-mixed livestocks, and the use of chemical inputs and concentrated feed, most of them imported.
9. For instance, this zone contributed 73 percent of sesame seed and 40 percent of the rice to the national agricultural production in 1974.
10. See: Josefina de Hernandez: "La implantacion Capitalista por Sustitucion de Anteriores Formas Productivas en la Agricultura Venezolana". Addressing to the XIth World Congress of Sociology. New Delhi, India. August 18-22nd, 1986.
11. By this time there were settled colonists from 27 different countries. The largest groups were Italians, Germans and Spaniards.

[illegible]

[illegible]

[illegible]

[illegible]

[illegible]

[illegible]

[illegible]

[illegible]

[illegible]

[illegible]

[illegible]

PART THREE

AGRICULTURE AND RURAL DEVELOPMENT

10

SMALL FARMERS AND FOOD PRODUCTION IN SOUTH BRAZIL
A Way Out of Imposed Technology

VILMA FIGUEIREDO

Introduction

The most recent and strong capitalist expansion in Brazilian agriculture (from the mid 60's to the early 80's) took place within an authoritarian political framework. Military governments replaced one another with no labour participation (urban or rural) in the public policy-making process. The exclusion of popular forces from the political decision-making arena resulted in an extreme concentration of economic power and wealth. One of the most fundamental consequences of this situation was the specialization of small farmers in agricultural products for export. In the south of Brazil small farmers abandoned the production of food for the domestic market as they increased technological inputs in their establishments. Though small farmers were aware of the negative effects of monocultural specialization, they also recognized the strategic importance of an adequate use of modern technology. This view and the concrete demands related to it have been expressed by rural labour unions, especially in the state of Rio Grande do Sul. Today the Brazilian democratic reconstruction seems to have created the conditions for satisfaction of these demands, primarily through the process of agrarian reform.

Concentration, Marginalization and the Modernization Process During the Years of Authoritarianism

The recent agricultural modernization in Brazil favoured the reproduction of capitalism, concentration of means of production, and marginalization of labour. However, concentration and marginalization stemmed more from the specific political decisions made within the Brazilian authoritarian system than from the general process of capitalist expansion.

Perhaps the most striking evidence of the specific manner in which such agricultural modernization developed is profit concentration. One percent of the richest rural producers controlled 10.5 percent of the total rural earnings in 1970 and 29.3 percent in 1980, while the poorest 50 percent controlled only 22.4 percent of such earnings in 1970 and 14.9 percent in 1980.

After 1964 the political posture of the Brazilian central government favoured international capital. It also addressed the agrarian question through technological modernization without paying attention to the specific and complex manner in which relations of production have been historically constructed in Brazil. In essence, the governmental strategy consisted of favouring the traditionally privileged classes within a dependent capitalist context. The vicious cycle of dependency which is synthesized in authoritarian capitalism became manifest from the mid 60's onward. The rejection of a popular ideology as well as the practice of consulting the people originated a style of government functional to the interests of international capital and to the adoption of "technological packages". Accordingly, in the name of greater "scientific efficiency", hidden interests committed to specific views of modernization were favoured.

The answer to demands for food, energy, and raw materials was found in increased agricultural consumption of modern inputs. Such inputs were originally produced locally by multinational corporations (as in the case of machinery and agricultural equipment, insecticides and fertilizers) or by state enterprises [the EMBRAPA System (Brazilian Enterprise of Agricultural Research)]. Such corporations were particularly concerned with adapting "technological packages" to national and local conditions. This strategy was made possible primarily through a governmental credit policy which favoured large landowners. The set of public policies aimed at modernizing agriculture in the name of greater productive efficiency generated an acceleration of the processes of land and income concentration in the rural milieu. However, and contrary to government propaganda, the process of modernization did not create economic efficiency, as productivity continued to be very low. Only a few products, such as soybeans, reached high levels of productivity, and these had only recently been introduced to Brazilian agriculture and were thus not traditionally cultivated by large landowners.

It can be argued, then, that agricultural modernization introduced in the last decades cannot be justified in terms of economic effectiveness. It reflects, rather, the private interests of large landowners and industrial conglomerates.

Some Consequences of Modernization on Small Farmers

Restructuring of the small farming sector accompanied modernization. The viability of autonomous small farms was reduced by three phenomena: proletarianization of farmers, marginalization of small farm family labour, and capitalization and technical modernization of small holdings.

The proletarianization of farmers resulted from the expansion of capitalist farms, consequent ability of these farms to invest and control trade, and the technical assistance they received which diminished the capacity of small farms to compete. Accordingly, petty commodity producers were forced either to give up farming or become agricultural wage workers. Both phenomena fostered a subsequent process of rural exodus (Moreira, 1981). In 1970 the urban population exceeded the rural population for the first time (55.9 percent). This exodus continued, and in 1980 67.7 percent of the total Brazilian population were urban residents. The increased proletarianization of petty commodity producers has had particularly serious consequences for the "boia-fria," i.e. temporary wage labourers with no employment alternatives in or out of farming. The ability of these individuals to find work in agriculture has decreased with a simultaneous increase in their precarious socio-economic conditions.

The marginalization and simultaneous subordination of small farm family labour to industrial capital occurred within a double movement of circulation where technical assistance played a fundamental role. Such circulation involved the exchange of inputs from the industrial sector to agriculture, and agriculture's raw materials to industry (CPDA, 1979). This process was selective as it only involved farmers who were able to adopt modern techniques and pertained to the production of several commodities of high commercial value such as tobacco, swine, poultry, and grapes.

There were specific structural and contingent situations which allowed the capitalization of small farmers. One example is the case of small landowners who profited from public incentives to change the character of their business. They began hiring wage labour and expanded the size of their farms through purchasing and/or renting new land. Instances of this pattern can be found in Cruzeiro do Sul and Rio Grande do Sul, where small farmers who benefited from soyabean incentives expanded their agricultural business. This is also the case of "Southern farmers" who moved to the Federal District as tenants of public land within a governmental settlement programme (Figueiredo, 1979). In addition, the process of capitalization occurred within cooperatives which provided farmers with the means to achieve a greater volume of production and to enhance the management of their farms.

In regard to technical modernization, small farmers adjusted their production to the terms of contract imposed by input providing firms. This phenomenon contributed to a reduction in the area devoted to the cultivation of products of lower commercial value and/or food for subsistence. Small farmers who capitalized their establishments and participated in public programmes which linked credit to technological innovation also developed monocultural productions. It follows that small farmers discontinued production for the domestic market in proportion to the increase in use of modern technology on their farms. This is a general tendency in the south of the country, especially among soybean producers.

Together with several case studies, census data provide additional information concerning this process. Between 1975 and 1980 the number of farms increased from 4,993,252 to 5,167,578 and the cultivated land from 323,896,082 to 369,587,872 hectares. The land employed in the cultivation of staple crops such as beans, manioc, and corn also grew from 323,615,963 hectares in 1975 to 38,687,995 hectares in 1980. In the North and the Northeast the number of farms and the land they occupy increased between 1975 and 1980. In the Southeast, Center-West and South the number of establishments decreased, while land employed for the cultivation of staple crops grew. This suggests that in the latter regions the cultivation of staple crops was developed through the use of more advanced techniques, yet was concentrated in a smaller number of farms.

In essence, then, while the restructuring of the small farm sector reduced the ability of small farms to compete, it should not be equated with their elimination, as they persist, though in decreasing numbers. Furthermore, they still play an important role in agriculture as in the production of rice and beans, though some studies show that even in such cultures capitalist production is expanding (Borges and Servilha, 1978).

Rio Grande Do Sul

Of the 50 municipalities (municipios) in which major land surfaces are occupied with staple crops, 19 are located in the state of Rio Grande do Sul. In this state, the number of farms and the land they occupy increased in all size categories with the exception of the group of small farms in the category ranging from 10 to 100 hectares. This group of small farms, however, has potential for capitalization though it also contains small farms which are in the process of being expelled from commercial agriculture. It is interesting to note that the number of farms producing staple crops in this group decreased, but the land devoted to the cultivation of these crops increased.

Among the four staple crops mentioned above, manioc is the only one that is not prominent in the South in terms of area and volume produced. The Northeast accounts for more than half of its production, yet has the lowest level of productivity. In the case of soybeans, corn, and beans, higher levels of production are recorded in the South, which also has high average rates of productivity. The South is responsible for almost all of the country's soybean production, as shown in table 10.1.

The performance of the South with respect to the above mentioned products is due to a number of factors. Among those deserving mention are the existence of rural colonies and the absence of slavery. Both factors made possible an intensive absorption of new technologies not only by large and medium establishments, but also by small farms. Given this particular situation, it was in the south, and specially in the state of Rio Grande do Sul, that small farmers have become politically active against the strategy of "modernization" carried out by the government. Their action has been based on their experience with modern agriculture and is centred around political alternatives that are compatible with a democratic regime.

Adequate Technology and Food Production:

The Point of view of Southern Small Farmers

Data gathered through case studies in Rio Grande do Sul (Figueiredo and Araudo, 1984) show that the use of modern technology (equipment,

TABLE 10.1: **Cultivated Area Quantity and Average Productivity Per Region, 1975—1980 (Corn, Soybean, Bean, Manioc)**

Regions	Cultivated Area (ha)		Quantity (Ton)		Average Productivity (Kg/Hectares)	
	1975	1980	1975	1980	1975	1980
(Corn)						
Total	10,854,687	11,451,297	16,334,516	20,372,072	1,504	1,779
North	108,368	174,967	116,147	221,931	1,071	1,268
Northeast	2,497,651	2,193,108	1,614,457	830,460	646	378
Southeast	2,964,018	2,940,532	4,682,865	5,595,565	1,579	1,902
South	4,396,538	5,145,339	8,307,755	11,638,995	1,889	2,262
Centre-West	888,112	997,351	1,613,292	2,085,121	1,816	2,090
(Soybean)						
Total	5,824,492	8,774,023	9,893,008	15,155,804	1,698	1,727
North						
Northeast	757	1,986	709	2,320	936	1,168
Southeast	467,197	723,156	765,656	1,388,600	1,638	1,920
South	5,106,658	6,918,703	8,780,627	11,856,126	1,719	1,713
Centre-West	249,880	1,130,178	346,016	1,908,758	1,384	1,688

(Contd.)

TABLE 10.1 *(Contd.)*

Regions	Cultivated Area (ha)		Quantity (Ton)		Average Productivity (Kg/Hectares)	
	1975	1980	1975	1980	1975	1980
(Beans)						
Total	4,145,916	4,643,409	2,282,466	1,968,165	550	423
North	35,193	63,306	38,146	36,598	1,083	578
Northeast	1,791,751	1,791,329	707,675	501,283	394	279
Southeast	900,321	1,221,193	445,270	672,143	494	550
South	1,140,918	1,258,993	932,899	662,600	817	526
Centre-West	277,733	308,588	158,476	95,541	570	309
(Manioc)						
Total	2,041,416	2,015,857	26,117,614	23,465,649	12,793	11,640
North	145,041	207,942	1,742,685	2,609,615	12,015	12,549
Northeast	1,107,117	1,293,021	12,562,605	13,324,339	11,347	10,304
Southeast	245,071	194,243	3,917,567	3,004,202	15,985	15,466
South	451,805	260,916	6,548,683	3,622,136	14,494	13,882
Centre-West	92,382	59,735	1,346,074	905,357	14,570	15,156

Source: FIBGE (Brazilian Institute of Geography and Statistics). Agricultural Production per Municipalities (1975-1980).

insecticides, fertilizers, developed seed) is familiar to small farmers affiliated with rural labour unions. Such farmers have criticized the way in which new technology is produced and consumed in Brazil. One central tenet of this position is the rejection of the exaggerated use of technology and the related development of monoculture.

These farmers do not deny that modern technology may increase productivity, but they maintain that productivity should not be the only criterion for the development and diffusion of technology. The pursuit of such a goal might signify the dependency of small farmers to banking and corporation groups, a risk of impoverishment, and the loss of land. The central issue, then, is the increased use of modern technologies vis-a-vis the neglect of existing alternative technologies. This situation, in the view of small farmers, is due to the prevailing Brazilian model of development, in which decisions are imposed from outside and from top to bottom. They conclude that an alternative must be sought in which the development of technology is considered in terms of the interests of farmers, consumers, wage workers, and the environment. Furthermore, in this alternative strategy, the knowledge of technicians must be combined with the experience and knowledge of farmers.

The adequate use of technologies, in the small farmers' view, will contribute to augment their income and also make possible the preservation of the environment and the production of nutritionally sound food items. The moderate use of modern technology combined with traditional techniques permits a cost/benefit ratio which is also favourable to the small farmer because it makes crop diversification possible. A reduction in the exaggerated use of machinery and agrotoxics will prevent soil erosions, plant and animal destruction, and possible damage to the health of workers and consumers.

In Rio Grande do Sul, small farms have been active through their rural labour unions in informing the population. There have been some successes, though these have been partial and temporary. The control of the use of agrotoxics through the mandatory requirement of an agronomic prescription is a case in point. However, the greatest success of rural labour unions had been the creation of a set of articulated demands which now are considered to be relevant by different segments of the local population and several political parties (Figueiredo, Teixeira, Araudo, 1985). It is obvious that the implementation of policies in accordance to the wishes of rural labour unions will depend upon the strengthening of democracy in the country and on the increased involvement of all segments of society in the political decision-making process.

The agrarian reform is an example of the problems associated with the process of democratization of society. The reform has been disrupted by the action of large landowners and other groups associated with them. However, the land reform has also been successful in some cases where important sectors of the population have been engaged in it. According to the reform, decisions concerning the expropriation and awarding of land must be carried out locally, state by state. This situation makes it difficult or even impossible to implement the reform in states where latifundists have greater economic and political power. However, where it has been implemented, the agrarian reform has not been a mere land distribution but has fostered the development of policies supporting small production both at the individual and collective levels. The small farmers of Rio Grande do Sul created the conditions necessary for the implementation of the agrarian reform within such a framework. Specifically, they made possible the participation in the decision-making process of groups once systematically excluded from it. Moreover, they were instrumental in redefining the country's parameters of economic development. If such a posture becomes dominant, food production in the South will significantly increase and production will be diversified. Also, if southern small farmers' demands for a controlled use of technology are satisfied, the quality of food products will be enhanced. This objective is clearly relevant not only for small farmers, but also for the entire Brazilian population.

References

Araudo, Caetano Ernesto P. - *Cooperativismo e Acumulacao: Um Estudo de Caso*. Dissertacao de Mestrado. UnB, 1980.

Borges, M. e Servilha, Waldemar - Producao Agricola de Alimentos no Brasil: Estudo sobre as Culturas do Arroz e Feijao (Mimeo), 1978.

CPDA - *Evolucao Recente e Situacao Atual da Agricultura Brasileira* Brasilia: Binagri, 1979

Figueiredo, - Vilma et alli - "A intensificacao da Agroempresa no DF" in *Reflexoes sobre a Agricultura Brasileira*. Rio: Paz e Terra, 1979.

Figueiredo, - Vilma e Caetano E. P. de Araudo - "Technologia Agropecuaria e Movimento Sindical de Trabalhadores Rurais". *Cadernos de Difusao de Technologia*. Vol. 1 no. 2, 1984.

Figueiredo, Vilma, Joao G. Teixeira e Caetano E. P. de Araudo - "Technologia Agropecuaria: O Parlamento e as Reivindicacoes dos Trabalhadores Rurais "*Serie Sociologia n 54*. SOL/UnB, 1985.

Moreira, Jose Roberto - "A Pequena Producao e a Composicao Organica do Capital". *Revista de Economia Politica*, Vol. 1 no. 3 Sao Paulo: Brasiliense, 1981.

[illegible] development of [illegible] segments of society in the political decision-making process.

The Brazilian reform is an example of the problems associated with the process of [illegible] of society. The reform was [illegible] by the action of large landowners and other groups associated with [illegible]. [illegible] has also been [illegible] in areas where important sections of the population have been engaged in [illegible] [illegible] are [illegible] the [illegible] of land [illegible] be carried out [illegible]. This [illegible] difficult [illegible] to implement the reform [illegible] where latifundias have [illegible] economic and political power. However, where unions [illegible] implemented the agrarian reform [illegible] been a more [illegible] has resulted in the development of [illegible] of the individual and collective level. The small [illegible] of the [illegible] created the conditions necessary for the [illegible] of the agrarian reform [illegible] political framework. [illegible] they made possible the participation in the decision-making process of groups [illegible]. Moreover, they were instrumental in [illegible] of economic development [illegible] [illegible] of small farmers but also [illegible].

[illegible]

[illegible]

[illegible]

[illegible]

[illegible]

[illegible]

[illegible]

[illegible]

11

THE DEVELOPMENT OF SELF-MANAGEMENT IN YUGOSLAV AGRICULTURE
The Case of Slovenia

ANA BARBIC

The Principal Features of Agriculture and Agricultural Policy in Socialist Yugoslavia

Agriculture within the National Policy of Development

The first and basic goal of socialist Yugoslavia was to develop industry. At the same time, agricultural development was neglected, partly due to the limited resources assigned to it and the lack of experts (Barbic, 1985). The result of this orientation was twofold. First, there was not enough food produced to satisfy the domestic demand and to export. Second, the neglect of agriculture and the rapid development of industry led to a rural exodus which was not accompanied by the growth of agricultural mechanization necessary to replace the outflow of labour.

Table 11.1: **Agricultural population in Yugoslavia**

Year	Yugoslavia	BiH	Monte Negro	Croatia	Macedonia	Slovenia	Serbia
1961	49.6	50.1	47.0	43.9	51.3	31.1	56.1
1971	38.2	40.0	35.0	32.3	39.9	20.4	44.0
1981	19.1	16.6	13.0	14.5	20.5	9.2	24.5

Source: Statisticki godisnjak SFRJ (Yugoslav statisticalok), Beograd: 1969, 1976, 1984.

The first five-year plan (1947–1951) failed to achieve its goal of reaching the level of agricultural production which had existed before the war. Operators of privately owned farms, who at the time cultivated 96 percent of all agricultural land, were forced to sell their agricultural products to the state in order to satisfy the basic needs of the non-agricultural population. Additionally, Stalin's criticism (especially strong in 1948 and 1949) indicating that Yugoslavia was avoiding the collectivization of agriculture forced the government to implement "working cooperatives" based on collective ownership of the land. Most of the farmers resisted this action, which was foreign to the Yugoslav tradition of cooperatives and relied on a model imposed from without. Since the rate of agricultural production did not increase, the forced collectivization was soon given up and by 1952 most of the "working cooperatives" were abandoned (*Kmetijsko zadruznistvo na Slovenskem*, 1982: 25) and publicly defined as failures (Kardelj, 1959: 7). However, the goal of a progressive socialization of the land was retained, for it was perceived as a cornerstone to higher agricultural production and rural development.

Despite the awareness concerning the importance of agriculture for overall national development, the pre-war food production level was reached only in 1956. However, production was still not great enough to satisfy the domestic demand. Great quantities of food were, thus, imported, especially wheat; and the trade balance was negative, as imports were twice as great as exports. The principal reason for the stagnation of agricultural production was insufficient investments. However, a policy of low consumption and low retail prices also kept the standard of living of the non-agricultural population low and aided accumulation of capital for the overall economic development (*kmetijsko zadruznistvo na Slovenskem*, 1982: 25). The actual neglect of the development of agriculture, the development of industry, and the urbanization process resulted in a reduction in the amount of agricultural land cultivated (Table 11.2).

Although larger collectively owned farms were slowly established, the majority of agricultural land still belonged to the family farms. According to a special law, farm families were not allowed to own more than 10 hectares of arable land situated in lowlands and not more than 20 hectares in hilly areas. The fact that private farmers could rent unlimited areas of arable land and cultivate it did not represent a solution. In fact, there was not enough land to be rented where a demand for land existed and, simultaneously, there were not enough farmers where such land was available.

Table 11.2: **The Ownership of Arable Land in Yugoslavia**

Year	Arable land in 000 hectares	Social Sector in %	Private Sector in %
1961	10.330	11.2	88.8
1965	10.280	13.7	86.3
1968	10.188	14.4	85.6
1970	10.153	14.7	85.3
1971	10.125	14.7	85.3
1972	10.092	14.9	85.1
1973	10.035	15.0	85.0
1974	10.007	15.1	84.9
1975	9.962	15.4	84.6
1977	9.950	15.7	84.3
1981	9.936	16.9	83.1
1982	9.892	16.9	83.1
1983	9.875	17.1	82.9

Source: Statisticki godisnjak SFRJ (Yugoslav statistical yearbook), Beograd: (1963, 1967, 1970, 1972, 1973, 1974, 1975, 1976, 1977, 1978, 1984).

Data show that in Slovenia in 1984, 31 percent of family farms cultivated an average 1.8 hectares of land, while 44 percent would have liked to cultivate more if they had the opportunity to do so. On the contrary, only 4.5 percent of farmers rented some of their land (on the average 1.5 hectares per farm). (Kovacic, 1984: 42-47).

Small family farms were and still are fragmented, thus many small plots of land. This situation has hampered agricultural production and limited the full utilization of agricultural machinery. However, family farms in Slovenia are not well equipped with agricultural machinery. Though in 1974, 71 percent of family farms owned one or more tractors, only 30 percent of them had different tractor attachments (Kovacic, 1984: 47-51).

Other factors have also contributed to the slow increase in agricultural production, (see Table 11.3) Among these, low levels of investment by family farms is perhaps the most important one (*Kmetijsko zadruznistvo na Slovenskem,* 1982: 25,26: Veselinov, 1984: 20).

Since the private agricultural sector has been viewed as a leftover of capitalism which will (sooner or later) be replaced by socialist agriculture, agricultural policies, particularly those concerning investments, have been oriented towards socially owned agricultural operations. It was believed that the larger these operations were, the more productive they would be. As a result of this investment policy,

Yugoslav agriculture today faces the problem of having socialist agricultural operations which are too large and privately owned family farms which are too small.

Table 11.3: **The rates of agricultural production in million din**

Year	National Product in Agriculture	Index (1970 = 100)
1957	31.567	80
1961	32.349	82
1970	39.188	100
1971	41.406	105
1972	40.469	103
1973	44.184	112
1974	46.627	118
1975	45.288	115
1976	48.412	123
1977	51.078	130
1978	48.213	123
1979	50.291	128
1980	51.018	130
1981	52.378	133
1982	56.293	143

Source: Statisticki godisnjak SFRJ (Yugoslav statistical yearbook), Beograd: 1978, 1982, 1983, 1984.

Large Socialist Farms

Socialist operations, especially pig, cattle, and poultry farms, are not only inefficient from organizational and productive points of view, but also in terms of their ecological effects on the environment and the reduction of soil fertility (Zagozen et al.,1986: 11). Accordingly, they employ a large quantity of energy, industrial, and chemical inputs which reduce significantly their economic and social efficiency. At the same time, the enormous quantities of liquid manure from pig and cattle farms pollute subterranean water, endangering the supply of drinking water. Moreover, the ecological impact of large farms damages the soil due to the usage of heavy machines and too narrow a rotation circle. Consequently, this land is subjected to water and wind erosion, to the intensified washing out of plant nourishment, as well as the loss of nitrogen to the air (Suncic et al.,1986), which must be added again to the soil by chemical fertilizers. This vicious cycle needs to be broken by restoring the natural equilibrium between animal and plant production.

Small Family Holdings

After the liberation (1945) the goal of socialist revolution, "Land To Those Who Cultivate It," was put into practice by expropriating landlords and by giving portions of their land directly to small farmers and landless farm workers. In order to provide small and middle-size farmers better chances of remaining in agriculture, their pre-war debts were abolished. Furthermore, the creation of "working cooperatives" stimulated the development of the socialist agricultural sector through the acquisition of more land and the use of available techniques (machinery, new technologies, etc.).

The private agricultural sector was not able to keep up with the development of the socialist sector primarily because of the less favourable investment policy and the small size of farms. As a consequence, farm labour was forced to seek employment in other branches of the economy. However, the decrease in the agricultural population did not automatically lead to an increase in the size of family holdings, for most full-time farms became part-time ones. Many of them cultivated their small pieces of land for their own consumption. Most of these farms had up to 1 ha of arable land only. In 1984 such households owned 15 percent of the privately owned arable land in Slovenia, out of a total of 85 percent.

If a farm is defined as an operation with at least 1 ha of arable land, then the average Slovenian farm has 4.3 ha of arable land (Kovacic, 1986). Such a small size cannot guarantee an adequate income for the farm family. Therefore, many farm family members have opted for off-farm activities. In Slovenia only 28 percent of all farmers are full-time (husband, wife and offspring work only on the farm), 59 percent are part-time farmers of the first type (at least one of the family members makes his/her living solely from agriculture) and 13 percent are part-time farmers of the second type (all of the family members are employed in a nonfarming activity and farm in their free time (Barbic et al., 1984: 14). If classified according to the sources and amount of income drived from agriculture and non-agricultural activities, the percentages of full-time and part-time farms are similar (Kovacic, 1986):

- Full-time farms: 26
- part-time farms (at least half of the income comes from farming): 23

– complementary farming: 23
– farm households with no active income (the aged): 8

The tendency is one of a shift from full-time farming to part-time farming and from part-time to complementary farming, with complementary farming showing the greatest degree of stability. No contrary trends have been observed to be significant.

In the process of transition from full-time to part-time farming women tend to remain on the farm while men tend to take an off-farm job and make farming their secondary activity. The proportion of women in agriculture in Slovenia is, according to some authors, 56 percent (Gilha, 1983) and, according to others (Kovacic, 1984) 60 percent of the active agricultural labour force.

Table 11.4: Active farm population in Slovenia by age categories and sex in 1981)

Age Category	Total Number of Farm Population	Men %	Women %	Total %
10-14	11	63.6	36.4	100.0
15-19	2,542	60.2	39.8	100.0
20-27	6,776	53.5	46.5	100.0
28-44	20,070	40.8	59.2	100.0
45-49	14,684	41.4	58.6	100.0
50-59	32,807	41.9	58.1	100.0
60-64	9,393	43.8	56.2	100.0
over 65	23,837	53.7	46.3	100.0
Total	117,120	45.2	54.8	100.0

Since the structure of the private agricultural sector is quite stable, the number and size of small farms is not expected to change in the immediate future. However, the growing costs of food and the generally decreasing standard of living will most likely increase the amount of part-time farming. Accordingly, strategies must be developed for the production of cheaper agricultural goods, one of the most promising of which might be cooperation among farmers.

Self-management in the Agricultural Sector

Self-management in the socialist sector of agriculture operates in the same way as in other organizations of associated labour. The

Associated Labour Act (1976) regulates the status of individual farmers and the conditions for the pooling of labour, land, and implements into agricultural cooperatives or into organizations of associated labour. It also indicates that associated farmers have basically the same status, rights, and obligations as workers in similar situations in other economic sectors, and farmers can exercise their self-management rights through farmers' cooperatives when membership is established.

Agricultural Cooperatives – The Tradition of Slovenian Farmers

Cooperation among Slovenian farmers is an established tradition dating back to the beginning of the 19th century, when farmers were forced to cooperate in order to survive. In the middle of the last century, farmers' credit cooperatives were established with the aim of protecting farmers against high interest rates charged by private bankers. Credit cooperatives were one of the most important links in the structure of the pre-war cooperatives. However, some other types of cooperatives were established as well. Among them agricultural, cattle-breeding, trade machine, fruit and wine, milk, and wood-processing cooperatives were the most popular. In 1940 there were 1,677 different kinds of cooperatives with 257,522 members. (*Kmetijsko zadruznistvo na Slovenskem, 1982:* 12, 17).

Towards the end of World War II (1944) the National Liberation Council prepared the guidelines for the establishment of cooperatives and, immediately after the war, some specialized cooperatives (milk, wine, machine, waterpiping, etc.) were formed. However, in 1947 general agricultural cooperatives began incorporating already existing specialized coops. In the first part of 1948 there were 1,151 agricultural cooperatives – almost one for every village. Since they were too small to operate efficiently, a process of consolidation was initiated and the number of cooperatives decreased. Many of them began cooperating with the socialist agricultural sector and food-processing industries. Others changed into new organizations in which cooperation with farmers was just a small part of their activities. After the Associated Labour Act was passed in 1976, cooperatives began to be organized in new forms called Composed Organizations of Associated Labour (COAL). According to this act, associated farmers (the members of an agricultural cooperative) have the same status, rights, and obligations as workers employed in associated labour organizations in other sectors. Farmers and other members of a farm household are free to join an agricultural cooperative or other forms of farmers' associations. Farmers keep the ownership of the land and other productive means they bring into the cooperative.

The Structure and Activities of Agricultural Cooperatives

If a cooperative is too large to perform its socio-economic and self-management functions efficiently, it can be divided into smaller components which are called basic cooperative organizations. The members of a basic cooperative organization can form agricultural units, if they are farmers, or work in units if they are workers of the cooperative.

If an agricultural cooperative develops additional activities such as food-processing or any other industrial type of activity related to agriculture, its workers can organize themselves into a basic organization of associated labour.

To become a member of an agricultural cooperative one needs neither to own agricultural land nor be a full-time farmer. The only condition for membership is to work on a farm which sells all or a part of its agricultural production to the cooperative (*Kmetijsko zadruznistvo na Slovenskem*, 1982: 45).

The principal purpose of an agricultural cooperative is to create income on the basis of pooling labour, land, instruments of labour, and/or other resources of farmers (Associated Labour Act, paragraph 31). The development of income is stimulated through the continuous expansion of market production and stronger ties with food-processing industries.

It must be noted that the amount of income distributed to the associated farmers is rather small. Even worse, a portion of their gain is used to pay for production and marketing expenses. Thus, only about 3 percent of the cooperatives' income is actually distributed to the farmers. It has been observed that the slow development of self-management within cooperatives and the low interest of farmers in taking part in it is, at least partly, due to the rather slow development of income relations within agricultural cooperatives (Kmetijsko zadruznistvo na Slovenskem, 1982: 46).

Agricultural cooperatives may associate in a composite agricultural cooperative or with other basic organizations of associated labour into composed associated labour organizations. Within the agricultural cooperatives farmers also associate in production communities, the most popular in Slovenia in 1981 being the following ones.

The Organizational Scheme of Agricultural Cooperatives.

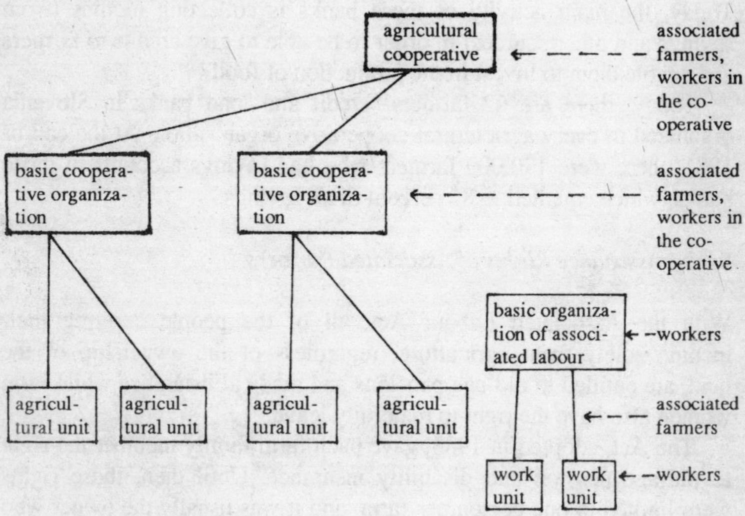

	Number of Communities	Number of members
- Communities with complete mechanization lines	3,470	10,660
- Communities with common use of one or several agricultural machines	662	2,559
- Milk-production communities	628	23,590
- Pasture communities	118	2,173
- Melioration communities	57	2,583
- Other production communities	606	5,347
- Total number of production communities	5,241	46,919

Source: Kmetijsko zadruznistvo na Slovenskem, 1982: 47.

It should be noted that the total number of Slovenian farms which own 1 ha or more of arable land is 105,690 (Kovacic, 1986).

The members of production communities pool together funds for buying agricultural machinery or erecting common buildings (stables, silos, etc.). In production communities there are also usually better conditions for the introduction of new technologies. As a rule, production communities are the strongest parts of agricultural cooperatives.

Another method for mutual collaboration and help among farmers and other rural dwellers has been farmers' credit and loan banks. Today, the main activity of these banks is collecting monies (from savings and other sources) in order to be able to give credits to farmers and enable them to invest in the production of food.

Today, there are 62 farmers' credit and loan banks in Slovenia organized in every agricultural cooperative organization. At the end of 1981, there were 170,000 farmers who had savings accounts in those banks, which equalled to 85 percent of all savings.

Social Assistance Rights of Associated Farmers

With the Associated Labour Act, all of the people deriving their income solely from agriculture, regardless of the ownership of the land, are entitled to old-age pensions and medical insurance while farm women also have the right to maternity leave.

The Act, adopted in 1983, gave each farm family member the right to old-age pension and disability insurance. Until then, those rights were limited to one person per farm, and it was usually the owner who was entitled to this right. The same was true for medical insurance for farm family members. Slovenia is the only Yugoslav republic which has so-called national health insurance, which gives each citizen the right to unlimited medical treatment and requires the payment of dues. The sum paid by farmers is calculated on the basis of estimated farm income per hectare. In order to gain both kinds of social assistance a member must fulfill two conditions. First, an individual must be a member of an agricultural cooperative and, second, he/she must contribute regularly to social security. However, the amount of social security contributions paid by the farmers is a source of controversy. Because of the socially controlled prices of agricultural products, farmers' income is lower than that which could be obtained if market prices were employed, yet farmers must contribute to social security funds at the same rate as all other workers. It follows that farmers are forced to gain less and pay proportionally more than other workers. One result of this situation is a decline in the number of associated farmers.

Table 11.5: **The Number of Associated Farmers in 1984 and 1985, according to sex**

	1984	1985
-Male	74.1	72.2
-Female	25.9	27.8
%	100	100
N	(89,126)	(78,972)

Source: Zadruzna zveza Slovenije. XIII. redni obcnizbor. Ljubljana, 7. 41986.

There are several other reasons for the decline in the number of associated farmers in recent years. First, many old farmers took advantage of the availability of old-age pension insurance and retired. Second, the economic conditions of agriculture have deteriorated recently, motivating the exit of a number of farmers from the sector. Finally, the politically imposed price disproportions between agricultural products and inputs is too harsh a condition to bear for many agriculturists, who as a consequence have changed jobs.

Perhaps the greatest advancement in social security for farmers was achieved in 1982, when the self-management agreement act concerning maternity leave was passed. According to this agreement farm workers have the right to maternity leave, which consists of one year of pay and the cost of all expenses associated with supporting the newborn baby. It can be expected that this new benefit will motivate some young women to stay on the farm and contribute toward a reduction and/or reversal of the exodus of young individuals from agriculture.

Farmers' Participation in Self-management and in the Decision-making Process

In order to participate in the decision-making processes at least two conditions must be fulfilled. First, there must be communication channels (formal and informal) available and, second, the advantages of such participation must be recognized by the people. In other words, in order to participate, people must play their role in improving their socioeconomic conditions through their own effort and actions.

As far as channels of communication are concerned, the political system of socialist self-management in Yugoslavia enables each citizen to participate in the decision-making process through the delegation system and other forms of organization.

The members of a cooperative (associated farmers and workers in the cooperative) exercise their socio-economic and self-management rights by expressing their individual views by voting in meetings and referendums.

The self-management organ of a cooperative is a cooperative council composed of the delegates of farmers and delegates of workers in the cooperative. More than half of the council delegates must be farmers. The executive organ of a cooperative is the executive council which is formed in the same fashion as the cooperative council. The latter nominates the director of the cooperative for a four year term. Within an agricultural cooperative or within its organizational units (basic cooperative organizations, basic organizations of associated labour), farmers and workers discuss proposals which are implemented according to decisions reached by majority vote. These organizations also act as candidature conferences and voting units where delegates for the cooperative council are elected. In this respect, they represent the delegation basis for elected delegates. Agricultural and working units also make all of the decisions related to their own affairs.

Self-management in agricultural savings and loan banks is also based on the delegation system. The assembly of the delegates (more than half of them have to be savers) makes decisions about the general policy of the bank, while its executive committee is responsible for implementing them and supervising all of other duties of the bank as a commercially efficient organization.

Farmers' Participation in Self-management

The data collected by interviewing 2,052 farm-heads (men and women) show that the number of farm heads who have at sometime been members of self-management bodies of agricultural cooperatives or local communities and communes is rather high.

One fifth (19.7 percent) of male respondents participated in self-management bodies of organizations of associated farmers, while only 13 percent of rural women did the same. This datum indicates that compared with farmers, farm women are under-represented in self-management bodies, particularly considering the fact that women in Slovenia represent one half (54.7 percent) of the population active in agriculture (A. Barbic, M. Rupena, V. Veslic, 1985; 11).

Besides this disproportion, there is little difference in the opinions of male and female farm heads concerning the adequacy of representation of men and women in decision-making bodies. Considering the low representation of farm women in the self-management bodies and delegations, it has been hypothesized that women are not satisfied with the existing situation. Such ideas have been partially confirmed by the data.

Number of farm-heads who have been (are) members of self-management bodies on different levels

		Percent
(a)	Within the agricultural cooperative	
	— members of self-management bodies of cooperative or basic cooperative organizations	17.9
	— members of self-management bodies of farmers' savings and credit banks	7.9
(b)	Within the work organization (part-time farmers)	
	— members of self-management bodies	10.2
	— members of the work organization delegations to the societies of self-interest and others	4.3
(c)	Within the local community and commune	
	— members of local community delegation societies of self-interest and others	18.5
	— presidents or members of LC council, its organs or committees	14.4
	— members of organs or committees of the commune assembly	5.6

Table 11.6: **Farm-heads' opinions about the adequacy of rural women's representation in self-management and delegate bodies.**

	Male	Female	Total
Does not know	26.2	33.4	38.1
Adequate	46.6	38.9	44.6
Not adequate	27.2	27.7	27.3
%	100.0	100.0	100.0
N	1,510	542	2,052

Since it is a fact that fewer women than men participate in self-management bodies, male and female farm-heads have been also asked to name the reasons for this situation.

The high percentage of "don't know's" can mean a lack of interest on the part of the respondents in the problem under discussion, but it can also indicate a certain degree of satisfaction with what has already been achieved in our society in the field of integration of women in the self-management and decision-making process.

The same reasons for lower participation of women in self-management organs are often mentioned to the same extent by men and women, except for the reason that "men want most of the functions". This reason is mentioned by more women (7.2 percent) than men (4.4 percent). But on the other hand, slightly more men dare to assume "important functions". The most important reason for this situation is, according to the majority of male and female farmers, "lack of time". Therefore it can be hypothesized that rural women are overburdened with farm domestic work. This situation is more pronounced in the case of rural women who hold an off-farm job, which means the rural women cannot find time for activities off the farm (public activities) or take over those "responsibilities" which are traditionally assumed to be performed by men.

Farmers' Membership in Socio-political Organizations

Socio-political organizations within the agricultural cooperatives and rural local communities can contribute considerably to the development of self-management relations. Unfortunately, farmers' presence within them is sporadic. Only 2.5 percent of the farm-heads are members of the league of Communists, the most influential socio-political organization. Accordingly, it may be precisely because of their low representation in the League of Communists that farmers cannot adequately influence agricultural and rural policies. To be sure, rural areas are also characterized by the existence of non-agricultural economic activities. It follows that development should be achieved with the simultaneous growth of the farming and non-farming sectors. However, there is no doubt that agriculture remains the basic economic activity of rural areas and should also remain as such in the future if the objective of self-sufficiency in food production is accomplished. This means that in rural communities farmers must have an adequate and equitable representation within socio-political organizations.

The extremely low membership in the League of Communists does not mean that farmers are not loyal citizens of socialist Yugoslavia, but only that the Communist Party has not found ways to attract their membership. A partial answer to this problem lies in the character of the national economic policy, which has not paid enough nor adequate attention to farmers and to agriculture. The relatively high number of farm heads who are members of the League of War Veterans proves that farmers actively joined the revolution, after which not enough effort was made to treat them adequately within the socialist system.

Table 11.7: **Membership of farm-heads in socio-political and some other social organizations**

	Percent
-Socialist Alliance	67.1
-Red Cross	37.1
-Workers Unions	32.2
-League of War Veterans	18.0
-Society of Youth Friends	4.4
-League of Youth	2.7
-League of Communists	2.5
N	(2,052)

Development of Agriculture Depends on Development of Self-management

What is the degree of influence farmers have on agricultural policy?

Table 11.8: **The degree of influence of different bodies on agricultural policy in the eyes of farm-heads**

The Degree of Influence	Republic Ministry for Agriculture	Commune Assembly	Agricultural Cooperative	Associated Farmers	Individual Farmers
-very low	9.2	5.8	3.4	10.7	66.4
-low	8.1	7.1	4.3	43.1	14.1
-average	18.0	16.6	34.2	19.7	7.1
-high	15.2	40.8	22.2	12.5	1.8
-very high	34.8	19.5	27.2	3.9	0.8
-don't know	14.8	10.3	8.7	9.9	9.8
%	100.0	100.0	100.0	100.0	100.0
N	(2,052)	(2,052)	(2,502)	(2,052)	(2,502)

Obviously, farmers are aware of their small influence on agricultural policy formation. They are also aware that lack of political power has been one of the factors responsible for the rapid migration of farm family members from agriculture and also from rural areas. Simultaneously, they feel that state intervention has been ultimately detrimental to farmers and agricultural development in general. In order to enhance the status of farmers and develop self- management, the ability of farmers to make decisions concerning their work should be acknowledged. A return to the market paralleled by the development of self-management is perhaps the only solution to guarantee the development of agriculture. At the same time, this is also the way to remove the social inequality existing between farmers and non-agricultural workers.

Such development, however, can be achieved only through the involvement of farmers, the consideration of their tradition and aspirations and by giving them an equal opportunity to speak about their rights in the political decision-making process.

Note

(1) The empirical data are taken from the study "Part-time farmers in the factor of the development of agriculture and the countryside". This study was carried out in 1982-83 in 13 of 65 communes of Slovenia on the basis of a representative sample of each of them. The principal sponsors of the study were communes who obtained the data for developmental planning.

References

Barbic, A. Idejna zamisel celovetega razvoja manj razvitega ruralnega obmocja (An outline scheme of integrated development of a less developed rural area). *IB revija za planiranje*. Ljubljana, no. 4. 1985, pp. 19-30.

Barbic, A. et al. *Mesane kmetije kot dejavnik razvoja kmetijstva in podezelja* (part-time farms as the factor of development of agriculture and countryside). Ljubljana, Biotehniska fakulteta, VTOZD za agronomijo, 1984.

Barbic, A. M. Rupena, V. Veselic, *Vloga kmetic v razvoju kmetijstva in ohranjanju podezelja* (The role of farm women in the development of agriculture and preservation of the countryside). Ljubljana, Biotehniska fakulteta, VTOZD za agronomijo, 1985.

Gliha, S. — Le se 9,1 odstotka kmetov (only 9.1 percent of farmers). *Sodobno kmetijstvo*. Ljubljana. no. 3, 1983.

Kardelj, E. *Problemi socialisticne politike na vasi* (The problems of the socialistic policy on the countryside). Ljubljana Cankarjeva zalozba, 1959.

Kmetijsko zadruznistvo na Slovenskem (Agricultural cooperatives in Slovenia). Ljubljana CZP Kmecki glas, 1982.

Kovacic, M. Kako s prestrukturiranjem kmetijstva? (How to restructure the agriculture?) *Nasi ragelegdi*. Ljubljana. no. 12. 20-6-1986.

Kovacic, M. Proizvodne znacilnosti kmetij (Production characteristics of family farms) in Barbic, A. et al. *Part-time farms as a factor of the development of agriculture and countryside*. Ljubljana Bitehniska fakulteta. VTOZD za agronomijo, 1984.

Kovacic, M. Radna snaga u individualnoj poljoprivredi SR Slovenije (Labour force in private agriculture of SR Slovenia). *Ekonomika poljoprivrede*. Beograd, no. 5, 1984.

Samoupravni sporazum o pravicah in dolznostzh zdruzenih kmetov v zvezi z uresnicevanjem pravice do porodniskega dopusta (Self-management agreement about the rights and duties of associated farmers related to the fulfilment of the rights to maternity leave). *Uradni list SR Slovenije*. Ljubljana, 1982.

Statisticki godisnjak SFR Jugoslavije (Yugoslav statistical yearbook). Beograd (1963–1984).

Suncic, F., J. Macek, M. Leskosek. Intenziviranje kmetijske pridelave ob ekoloskem ravnovesju (Intensification of agricultural production within the frame of ecological equilibrium). *Revija za razvoj*. Ljubljana, no. 1, 1986, pp. 12-16.

Veselinov, D. O., strategiji agrarnog razvoja jugoslavije (About the strategy of agricultural development of Yugoslavia). *Problemi socialistickos razvoja poljoprivrede i sela*. Ljubljana, 1984, pp. 18-22.

Vukcevic, R. Poljoprivreda u strategiji jugoslovenskof razvoja (agriculture within the strategy of the development of Yugoslavia). *Problemi socialistickos razvoja poljoprivrede i sela*. Jugoslovanski center za teorijo in prakso samoupravljanja Edvard Kardelj, Ljubljana, pp. 13-17.

Zagozen, F. et al. Prestrukturaranje preizvodnje z vidika racionalnega izkoriscanja proizvodnih danosti in pocenitve prozzvodnje hrane in surovin (Restructure of the production from the point of view of national exploitation of production resources and reduction the cost of food and raw material). *Znanstveno tehnoloski posvet biotehniskih ved*. Ljubljana, April 1986.

Zakon o zdruzenm delu (The Associated Labour Act) (1976), Ljubljana, CGP Delo.

12

EQUITY-ORIENTED RURAL DEVELOPMENT IN INDIA—ITS ORIGIN, SHAPING AND CONSEQUENCES

K.S. MANDAL

A review of the rural developmental programmes of independent India suggests that rural development in India has been in three distinct phases. In the 50s the rural developmental attempts resulted in a basic infrastructure for rural development in the country. In the 60s, the main concern of rural development was to achieve rapid growth through intensive agriculture. In the 70s rural developmental attempts were characterized by equity-orientation, and the same trend is continuing in the 80's. In brief, attempts at rural development in independent India have been marked by infrastructure building, growth, and equity phases.

The growth-oriented phase of rural development had resulted in a major imbalance in rural India and had the potential to disrupt the existing social order. However, the policy makers took a timely note of this trend and, as a corrective measure, growth-oriented strategy was replaced by equity-oriented strategy. This shift in state policy is very significant, as this brings out the equilibrium maintaining nature of the country. Thus, this paper attempts to examine how this equity-oriented rural development came into being in India; how the policy of equity-oriented rural development was shaped in the policy formulation and its operationalization stage; and finally, what are the consequences of this policy for the Indian society?

Several equity-oriented rural developmental programmes have been introduced in the country during the 70s and 80s.[1] Of these programmes, the Small Farmers Development Agency (SFDA) Programme was the pioneering one. This programme was very wide in coverage as this was a programme for small farmers, marginal farmers, and agricultural labourers – the core of the rural poor of India. There were other equity-oriented programmes which were in operation in some parts of the country for sometime, but SFDA was a national-level programme and was in operation for a long period spreading over the Fourth and Fifth Plan periods. In fact, the Integrated Rural Development Programme (and even the rural employment programmes), which characterize the rural development of the 80s, can be considered as an expansion and extension of the SFDA programme.[2] Because of this pivotal position of the SFDA programme in equity-oriented rural development, the present paper attempts to answer the above question in the context of SFDA programme.

Origin of Equity-Oriented Rural Development

In tracing the origin of equity-oriented rural development, the present section attempts to document how the SFDA programme was conceived and how the related policy was formulated.

The Reserve Bank of India operates as the monitor of the banking system in India. In this responsibility, in the year 1951-52, it conducted an investigation for assessing the rural credit situation and made recommendations. This was the first exercise of its kind in independent India. In the year 1966, the need for having a second look at the rural credit situation was felt by the Reserve Bank of India and thus a second All-India Rural Credit Review Committee was appointed by the Reserve Bank of India in July 1966. It is interesting to note that this step was initiated out of a feeling that there was uneven availability of institutional credit for agriculture in the country which was further aggravated by a larger credit requirement with the development of agriculture.[3] The Committee concluded among others that "the small farmers have not benefited in proportion either to their numbers or their needs from the various programmes of rural development which have been under implementation during the three plans (First, Second and Third Five Year Plans)... this is also broadly true in regard to the availability of institutional credit for agriculture."[4] It is this and similar findings of the Committee which resulted in the policy recommendation of the SFDA. Here, it is important to note how the committee arrived at these findings.

When the Committee started its work, the data of a comprehensive All India Rural Debt and Investment Survey, conducted in 1961-62, was available. Hence the Committee decided not to undertake any elaborate investigation on the lines of a usual Rural Credit Survey. Instead, the Committee pursued the following three steps.

First, the Committee decided to prepare a series of background papers on a number of relevant issues. Thus, 24 background papers and 17 working papers were prepared by the Secretariat of the Committee. Material for this was supplied by various quarters such as the Union Agricultural Credit Department, the State Bank of India, the Agricultural Refinance Corporation, the Food Corporation of India, the National Cooperative Development Corporation, the All-India Central Land Development Banks, etc.[5]

Second, the Committee made use of national scale data such as Census, the National Sample Survey and the All-India Rural Debt and Investment Survey.[6]

Third, the Committee decided to conduct a few field studies on a limited scale on related aspects of rural credit. Thus, two sets of field inquiries were organized simultaneously: one in respect of short-term credit in selected Intensive Agricultural District Programme areas and the other on long-term credit for financing investment in agriculture.[7] Table 12.1 presents the institutions and the type of studies undertaken by them in this regard.

Thus, we observe that various departments of the government, banking organizations, and research institutions were involved in providing the data or conducting study based on which the Committee took note of the fact that small farmers have not been duly benefited during the period under review. Table 12.1 also indicates that this finding was based on the studies of the impact of growth-oriented programmes like the Intensive Agricultural District Programme, High-Yielding Varieties Programme, and also on studies relating to the problem of small farmers.

Taking note of the fact that small farmers did not benefit duly from the developmental programmes, the Committee felt that steps should be undertaken to develop small farmers. The reasons given by the Committee for this can be summed up as follows[8] :

1. In terms of the number and area accounted by small farmers, they are significant enough from the point of view of agricultural production.
2. Keeping this big section of small farmers out of the ambit of developmental programmes will have far-reaching socio-economic consequences.

3. Increasing evidence shows that the productivity of small farmers is not inferior to that of large farmers.
4. The apprehension that new agricultural strategy may further aggravate already existing economic disparity among the cultivators.

Table 12.1: **Institutions that conducted studies and the type of studies undertaken for the All-India Rural Credit Review Committee**

	Institution	Type of study undertaken
1.	Division of Rural Surveys, Economic Department, Reserve Bank of India	(a) Evaluation of Intensive Agricultural District Programme and High-yielding Varieties Programme (b) Field studies on long-term credit (c) Assessment of the problems of the small farmer.
2.	Agro-economic Research Centres and Indian Institute of Management, Ahmedabad	Evaluation of High yielding Varieties Programme
3.	(a) Banking Division, Economic Department, Reserve Bank of India (b) Department of Economics and Statistical Research (c) Rural Credit Section, State Bank of India	Conducted studies based on operational data of banks on rural banking and agricultural credit in areas covered by IADP or other rural development programmes
4.	Agricultural Credit Dept. Reserve Bank of India	Assessment of the ability of the Cooperative Credit structure to support intensive agricultural programme
5.	Planning Commission	(a) Assessment of the problems of the small farmers (b) Evaluation of Intensive Agriculture District Programme

Source : Reserve Bank of India, Report of the All-India Rural Credit Review committee, RBI, Bombay (1969), pp. 96-97.

It is important to note here, that the Committee felt that depriving small farmers, who have great potential, might lead to undesirable consequences. Thus, this decision regarding small farmers' development was not taken under any ideological consideration – advancing the socialist goal of society was not any explicit consideration in initiating this policy.

That advancing a socialist goal was not a consideration for preaching small farmers development becomes clear when we examine the recommendation made by the Committee in this regard. In formulating the small farmers development programme, the committee confined its concern only to small farmers, excluding the marginal farmers and agricultural labourers whose conditions are far more critical. If the Committee were to be guided primarily by egalitarian considerations, a programme for the poorer marginal farmers and agricultural labourers would have been the priority. However, the Committee explained.

> "We concern ourselves here only with farmers whose business, including subsidiary activities like animal husbandry, is potentially viable and can in actual fact be rendered viable, if there is support in terms of irrigation facilities, surplus of inputs, services such as tractor-ploughing, technical guidance and so on. We leave out of consideration here–not because they are not important, but because they largely fall outside our terms of reference– other weaker sections of rural population such as sub-marginal farmers and agricultural labourers. What these require is a far-reaching programme of rehabilitation including but extending far beyond mere credit."[9]

Thus, it can be concluded that pragmatism and not egalitarian consideration was the prime factor in the initiation of equity-oriented rural development in India.

Shaping of Equity-oriented Rural Development

Having found that the benefit of developmental programmes and institutional credit did not duly flow to the small farmers, the Committee argued that there is a need of taking special measurers for them. The question is, what measures did the Committee suggest in this regard and how did they arrive at these? The present section focuses on this.

First, the Committee, in arguing for the need of making special arrangements for the small farmers, both for institutional credit and State assistance, drew upon the "Approach to Fourth Plan" which also made the similar point.

Second, the Committee argued that the problem of the small farmer and similar categories of farmers "required to be identified and dealt with in the specific context of their own environment.[10]

The Committee thus suggested the establishment of a pilot project – the Small Farmers Development Agency – in selected districts for assisting small farmers. This Agency should first help to identify the appropriate groups of farmers in its jurisdiction: (i) their special problems as producers, (ii) the means by which they can be helped to overcome their handicaps and render their farm economically viable and commercial, and (iii) the arrangements by which these means can in fact be provided. In carrying out the efforts on these lines, the proposed Agency will not extend credit itself, but will try to ensure that the cooperative credit institutions finance small cultivators under the programme.

It is interesting to note that the approach developed by the Agency outlined above has drawn upon a similar exercise of the bankers in the context of small-scale industry. A few years before this exercise, the State Bank and other commercial banks had worked out a scheme for the development of small-scale industries. This scheme aimed at coordination of efforts by different agencies, like supply of technical know-how by the Small Industries Service Institutes, supply of machinery as well as help in the marketing of products by the National Small Industries Corporation, supply of credit by the banks and so on. The suggestions regarding the SFDA programme are analogous to this.[11] As most of the members of this Committee were bankers, their familiarity with this earlier exercise of bankers might have contributed to the shaping of the approach for developing small farms.

It can be noted here that assessing the rural credit situation was the terms of reference of this Committee. But in the process of assessing the rural credit situation the committee found that both institutional credit and rural development programmes did not benefit the small farmer. Thus, the recommendation they made was beyond the sphere of credit. And the Committee justified, "if credit for such farmers is to be fruitful, it has to be provided as part of a wider effort to raise their economy to a surplus level."[12] Thus was the shift from credit to wider developmental activity.

It is important to note how the Committee's recommendations shaped into a policy. The Committee's terms of reference had made a specific mention about the Fourth Five Year Plan, and when the Committee was at work, the Fourth Plan was under preparation. For making use of this, the Committee decided to make interim recommendations on some important issues which it considered specially relevant at the stage of formulation of the Fourth Plan. Setting up of the Small Farmers Development Agencies in selected districts was one such recommendation, and it found its place in the Fourth Plan.

Subsequent to the recommendation of the All-India Rural Credit Review Committee, the Planning Commission, perhaps as a result of its commitment to egalitarianism, found it difficult to recommend a scheme that excluded the poorer sections of marginal farmers and agricultural labourers. Thus, under the Fourth Five Year Plan, in addition to the scheme for the small farmers, a separate programme was introduced for the marginal farmers and agricultural labourers, in line with the SFDA programme. The Centre-Sector Scheme for marginal farmers and agricultural labourers was devised by the Union Department of agriculture in consultation with the Planning Commission. The scheme aimed at assisting "the marginal farmers and landless agricultural labourers in view of the high priority attached in the Fourth Plan to measures specifically designed to enable the weaker sections of the rural population to benefit from the economic growth and development in the rural sector."[12]

Thus, two separate developmental programmes came into operation in the Fourth Plan period for the weaker sections of our society. The first one was the Small Farmers Development Agency (SFDA) programme, introduced to serve the interest of the small farmers. The second was the Marginal Farmers and Agricultural Labourers Agency (MFAL) programme for benefiting marginal farmers and agricultural labourers. Most of the projects under these two programmes completed one year of functioning in the year 1971-72.

The National Commission on Agriculture brought out its Interim Report on Reorientation of Programme for Small Farmers and Marginal Farmers and Agricultural Labourers Development Agency in August 1973. In this report, various recommendations were made regarding SFDA and MFAL programmes. The most important was that there should be one combined programme covering small farmers, marginal farmers, and agricultural labourers instead of two separate SFDA and MFAL programmes. Thus, it was decided that in the Fifth Five Year Plan the marginal farmer and agricultural labourers will be covered by the existing SFDA projects and small farmers in MFAL projects, and this combined programme will be known as the Small Farmers Development Agency programme.

The foregoing discussion points out how the SFDA programme

came into being, and how the programme was shaped in the policy formulation stage. However, a policy takes concrete shape in the process of operationalization. Thus, in the following section we will examine the shape taken by the SFDA programme in its operationalization. Here, we will confine our discussion only to two mechanisms that played an important role in the shaping of the SFDA programme. They also represent the spirit of the mutation that took place in the operationalization of the SFDA programme. These mechanisms are (i) implementation of an equity-oriented SFDA programme through a growth-oriented administrative set-up, and (ii) a group-centred approach in adapting modern technology.

Implementation of the equity-oriented programme through growth-oriented agricultural extension strategy

The area of operation of the Small Farmers Development Agency was a district. However, besides a monitoring office of the Agency located at the district headquarters, the SFDA did not have any separate administrative machinery. The SFDA project office of the district was to get its work done through the existing administrative set-up of the district, mainly the various developmental functionaries of the district. In this, the most crucial role was played by the Block Development office–the base of the developmental administration of the Indian government.

For an understanding of how the message of the SFDA programme reached the village and how the programme implementation was initiated, it is important to look into the operation of the agricultural extension of the developmental blocks. In most of the country the agricultural extension service of the developmental blocks is rendered through the Training and Visit System of Agricultural Extension proposed by the World Bank for spreading modern technology. Since the main mechanism for spreading the information about SFDA and activizing its programme implementation was the Training and Visit (TV) System of Extension Service, here we will briefly look into its operation.

Under the TV System of Extension Service a block is divided into several TV units, based on the total number of farm families and the number of families which a Village Level Worker (VLW) can reasonably be expected to cover. All the farm families under VLW's jurisdiction (which is a TV unit) are divided into eight groups of about equal size. From each group, the VLW is supposed to select about 10 percent of the farmers as "contact farmers" on whom he concentrates his efforts.

The key concept of the TV system of extension service is the concept of imitable contact farmers. According to this concept "The message of the extension service should be focused mainly on selected contact farmers who will assist in spreading the new practice to most farmers in the area quickly."[14] Thus, as the main purpose of the TV System of Extension Service had been to spread advance agricultural practices through the "contact farmers", the VLW naturally selects such farmers as "contact farmers", who can easily accept the new methods. This requires the farmer to be able to promptly meet the requirements of the new practice, and in addition, he should be able to bear the risk of trying a new agricultural practice. Both of these call for economic soundness of the farmer. The characteristics of the farmers primarily served by the extension worker is well documented by V. R. Gaikwad in the following lines.

> "It is a wellknown sociological fact that in an Indian village, by and large, most of the fertile land with irrigation facilities is owned by a handful of large joint families, generally belonging to the locally dominant caste. The extension worker...works mainly among this class of people, who traditionally enjoyed position of power and status in the village community."[15]

It is this class of "contact farmers" of the TV system of extension service, through whom attempts were made to spread modern technology in the 60s. Now, though under SFDA there was a change in the objective of the programme from growth-orientation to equity-orientation, the mechanism for delivering the programme remained the same. Thus, though the target group of the SFDA were small farmers, marginal farmers and agricultural labourers, delivery of the information about the programme and its activization took place mainly with this better-off "contact farmers" through the routine operation of the VLWs. And then it is this group which took the main advantage of the SFDA programme. When required they took the SFDA assistance as a group inclusive of the small and marginal farmers, but manipulating the main advantage of the programme in their favour. How this was done has been elaborated in the section that follows.

A Group-centred Approach in Adapting Modern Technology

In the 60s, the spread of modern technology in Indian agriculture remained confined only to the bigger farms. Small farmers were not able to take advantage of it. This increased inequality in rural India, and the SFDA came into being to remedy this situation. The SFDA's effort to make the small farmer economically viable mainly involved enabling them to take advantage of modern technology. Here the SFDA faced the following hurdles. The benefit of modern agricultural technology could not effectively accrue to an individual small farmer whose holding is also very fragmented. In fact, it has been pointed out that the individual farm-centred technology itself is fast reaching its limits. The potentials of modern technology can be harnessed effectively only through a group-centred approach.[16] For instance, technologies like run-off collection of water harvesting in non-irrigated areas need management of soil and water on watershed basis rather than individual farm basis. Efficient water management – a key factor in well-watered areas – also involves group action. Similarly, devices like integrated pest management, integrated nutrient supply, and improved post-harvest technology in their respective ways would demand collective endeavour on the part of the farmers in a contiguous area or in a watershed. Thus, the Ashok Mehta Committee has underlined the importance of "community action or group-action for taking full advantage of the emerging technologies and techniques most of which are either too large or too scattered or both to be absorbed by individuals acting on their own volition and isolation from their neighbours and potential cooperators."[17]

The SFDA, in its attempt to enable the small farmers in harnessing the benefit of modern technology, made a pioneering effort in introducing group-centred schemes. Here, we will illustrate the type of problem faced by the SFDA and how the SFDA fared in this regard. The illustration here is based on the author's observation of the SFDA schemes of Hooghly district of West Bengal.[18] However, the issues involved are similar under other SFDAs as well.

Our illustration here focuses on ground water irrigation and hiring of agricultural implements, the two most popular SFDA schemes. According to the SFDA statute, the maximum land of an SFDA farmer is five hectare, and as mostly the holdings of small and marginal farmers are highly fragmented, their holding at one stretch is of much smaller size. Even if a small farmer has his entire holding as a consolidated unit, its upper limit is five hectare, whereas a shallow tubewell along with a pumpset of a ground water irrigation system has a capacity of irrigating about 12 hectare of land of a composite watershed. Thus even for a small farmer with a consolidated holding of a maximum five hectare of land, owning such an irrigation system will be uneconomical for irrigating only his land, which will be underutilizing the capacity of this irrigation system. The same is true of many improved agricultural implements like tractors, power tillers, pesticide spraying machines, etc., the use of which will be economical in a much greater area of land than the individual holding of a small marginal farmer. The way out of this situation devised by the SFDA was the Community Irrigation Scheme, under which a cluster of shallow tubewells were sunk under SFDA subsidy for irrigating the land of a group of farmers of a composite watershed for ground water irrigation. Similarly, the SFDA provided subsidy for improved agricultural implements, service of which could be hired by the farmers of a composite watershed. These were owned and managed by a farmers group.

However, in this effort the SFDA faced the following problem. The holdings of the big farmers are contiguously located with small and marginal farmers. Thus a composite watershed of a Community Irrigation Scheme consisted of land belonging to big, small, and marginal farmers. Though the SFDA-aided Community Irrigation Scheme was meant for providing irrigation facilities to small and marginal farmers, it was not feasible to exclude the big farmer's land within one composite watershed of a Community Irrigation Scheme. Keeping this type of situation in view, the National Commission on Agriculture in its Interim Report on Reorientation of the SFDA programme suggested that, "Consolidation of holding should be accorded priority in the areas selected for the programme of development and special efforts made to bring the holdings of small and marginal farmers into compact block where preferential irrigation could be given to them through state-sponsored community wells for best results. In the absence of consolidation of holding and whenever a groundwater scheme is developed a group approach to the irrigation needs of small farmers would have to be considered by including, if necessary, the large farmers who might have their lands in the irrigation command *but taking care to see that the financial assistance in the shape of subsidy is made available only to small and marginal farmers* benefiting from the irrigation source" (emphasis added).[19]

It is interesting to see how this important guideline was operationalized.

In accordance with the above recommendation, the SFDA Hooghly stipulated the following rules regarding the Community Irrigation Scheme (CIS).[20]

First, that the subsidy to the CIS will be provided at the rate of 50 percent of the capital cost related to the ayacut possessed by small and marginal farmers.

Second, that the big farmers within the command area of the CIS will be charged water tax at the commercial rate, and small and marginal farmers will be charged water tax at the subsidized rate. A similar decision had been taken about the hiring charges of mechanized agricultural implements also.

But the above directives were violated in practice. First, for getting subsidy to the fullest extent, the land records were manipulated showing the big farmers in the group as small farmers. At the time of field work the record of 29 community irrigation schemes were available at the SFDA project office. It is interesting to note that in this record none of these 29 CIS reported of having land belonging to big farmers in its command area, whereas during the subsequent visits to the villages it was reported by the villagers as well as by officials that the command area of schemes does include land of big farmers.

Second, not a single CIS reported charging different rates of water taxes from big, small, and marginal farmers. Same was true about the hiring charges of the machineries. Since official records did not show the presence of any big farmers land in the command area of the scheme, it is but natural that there is only one rate for all farmers. This is so because usually it is the big farmers in the group who are at the helm of these groups' activities.

During the field work it was observed that a few active members, generally farmers with larger holdings, ran the entire show of this group's activity, and the rest of the members of the group played a very passive role. For instance, they were not aware of the SFDA subsidy for these schemes. They were not even aware of the fact that they were taking irrigation water and hiring services of agricultural machinery, as a member of the "cooperative society" that ran these schemes. Their role in this group activity can be characterized in the following way. They thought that as some people (generally the active members with bigger holdings) of their village have taken the initiative to get the Community Irrigation Scheme and other schemes sanctioned by the government, they were getting the advantage of those schemes

Their relation with the group was mainly one of buying irrigation water and hiring of agricultural machineries. When it was pointed out to them during discussions that they get these services as a member of a cooperative society which runs these schemes, the typical answer was that "it might be so, Mr 'so and so' (referring to the big farmer initiators) came and asked me to sign some papers which they said were required for getting water from the CI scheme and hiring implements and I signed." This exemplifies the nature of the participation of the majority of the members in most of the these groups. Obviously they do not participate in decisions of the group but are recipients of it. These groups were not run on the basis of "cooperation". They were rather an "operation by few and .... by others."

The above developments, in practice, have lopsided the priority of the SFDA programme, whereby it was the big farmers who for getting the benefit of modern technology with the SFDA subsidy took the initiative for getting SFDA schemes implemented and, because of "public or common good" delivery nature of technology, these schemes covered the small and marginal farmer of the composite watershed. However, the main advantages of the initial initiative of getting the schemes sanctioned and running the schemes accrued primarily to the big farmers. In this context the following observation is pertinent: "Not many bothered about equality or institutional reforms so long as technology performed the miracle. Now we have a situation in which the miracle-performing capacity of technology is itself likely to be seriously constrained by lack of suitable institutional arrangements. In such circumstances large farmers alone cannot get the benefit of technology (e.g. integrated pest management or improved water management) unless the small farmers in the same watershed are helped to participate in the process. The benefit of technology, in the changed circumstances, assumes the character of a "public or collective good" and nobody can be deprived of it as long as others (in a given watershed) have it. Thus *in a very indirect manner new technology at last offers a promise to poorly endowed farmer*" (emphasis mine).[21] The SFDA's role can be viewed in this perspective.

The SFDA, which was initiated as a corrective measure for the maladies of the earlier policy, in the process of its policy formulation was shaped by egalitarian considerations. However, the programme implementation process and collective good delivery nature of modern technology resulted a process where by the SFDA, in the process of benefiting the better-off farmers, apparently benefited the poorer farmers as well. However, the exact nature of this process is discussed in the sections that follow.

Consequences of Equity-oriented Rural Development

Understanding the consequences of the SFDA programme as an equity policy would involve the following steps. First, it would involve identifying the groups that have benefited from the SFDA programme and, second, an analysis of the extent of benefit to these groups. Obviously, the starting point of this analysis will be defining "benefit." Below we mention two points in this regard.

First, usually whenever a programme is administered to a person, often that person is considered as the sole beneficiary of the programme. Our contention here is that administering the services of a programme to a person does not necessarily mean that the very person got the actual benefit of the programme. For instance, it has been reported that there were cases where a big farmer, operating behind the scene, has made an agricultural labourer to get an agricultural implement with SFDA subsidy which eventually changed hands to the big farmer. Here, although officially the assistance has been administered to the agricultural labourer, the real beneficiary was the big farmer. Second, there could be a case where even availing of SFDA assistance by a person need not necessarily render benefit to the person concerned. For instance, SFDA assistance might enable a farmer to use modern technology resulting in increased production, but the high cost of cultivation and want of a remunerative price might result even in a negative income. The actual beneficiary in this case might be the agricultural input supplier whose payment (and profit) has been assured by an SFDA subsidy and bank loan credited to the farmer, the foodgrain merchant by increased (profitable) business, the agricultural labourer by additional employment and wage. Thus, a proper understanding of the nature of benefit from a programme involves an analysis of who ultimately benefits and to what extent by the administration of an assistance.

Second, the first point implies that besides the clients who are directly covered by a programme, the operation of a programme might indirectly generate benefit for some others as well. For instance, when the SFDA enables a farmer to undertake intensive cultivation, along with many other consequences, intensive cultivation generates employment for the agricultural labourers. This benefit, by way of additional employment and income, to the agricultural labourers can be considered an "indirect benefit" of the SFDA schemes. But as the SFDA did not administer any assistance to this group directly, they do not figure in the official statistics of SFDA beneficiaries.

Using the above conceptualization of benefit, we will now examine who have been benefited by the SFDA programme. In doing so, we will categorize the beneficiaries of SFDA under two categories: (a) Direct beneficiaries and (b) Indirect beneficiaries.

(a) *Direct beneficiaries:* Apparently the SFDA programme has benefited the households directly covered by its assistance. This group of SFDA (direct) beneficiaries consists of small farmers, marginal farmers and agricultural labourers. Among these direct beneficiaries, besides the abovementioned groups, there are also households having land more than the SFDA prescribed limit.

However, the extent to which these direct beneficiaries of the SFDA have been able to benefit out of the SFDA assistance is not very obvious. SFDA-sponsored expansion of modern technology enabled its recipient to undertake intensive cultivation which increased the production of these farms. However, during the field work, the SFDA beneficiaries repeatedly pointed out that often they did not get a remunerative price for their additional produce. At times, the beneficiaries also reported negative income, in spite of increased production. This phenomenon has been well documented in a study by Swamy and Gulati. They observed, though productivity per hectare in agriculture has increased in the 1970s.

> "A careful examination of economic conditions of cultivators and their farm income during the 1970s reveals... that the (net) incomes per hectare (or per holding) from cultivation have generally declined.... During the 1970s there is a dependable empirical evidence of a sharper rise in the prices of agricultural inputs, despite the cognizable quantum of subsidies granted by the government on certain important inputs, than those of agricultural outputs. For example, over the decade (1970-71 to 1980-81) fertilizer prices increased by 243 percent, pesticides by 329 percent and tractors by 290 percent, while prices of cereals have increased by 213 percent."[22]

Though the above study does not give an idea about the farm sizewise retrogression of income, it appears that among the abovementioned direct beneficiaries of SFDA, the better-off got the most–not only in terms of assistance from SFDA but also in terms of ultimate accrual of benefit out of the SFDA assistance.

(b) *Indirect beneficiaries:* The incidence of indirect benefit generation has added an important dimension to the above nature of benefit generation. These are the following four important types of indirect beneficiaries.

(1) The most important contribution of the SFDA programme was that it enabled its clients to adopt modern technology. It provided an infrastructural support to its beneficiaries by enabling them to get irrigation water, service of agricultural implements, chemical fertilizers, pesticides, etc. The SFDA in the process of providing these inputs to the farmer also assured a market for the traders of irrigation equipments, improved agricultural implements, chemical fertilizers, pesticides, HYV seeds, etc., by providing subsidy and a bank loan to its clients for buying these. Thus, this expansion of market under the SFDA assistance had indirectly contributed to the income of these traders.

(2) The SFDA programme, by contributing to the expansion of intensive cultivation, has also generated demand for agricultural labour. By generating additional employment and income for agricultural labourers it has benefited the agricultural labourers. Though the SFDA failed to directly deliver services to the agricultural labourers, it contributed to the agricultural labourers' income, as SFDA-sponsored intensive cultivation resulted in additional employment and wage hike for agricultural labourers.

Above we have identified the group that have been benefited, directly or indirectly, by the operation of the SFDA programme. No precise data is yet available regarding the extent of benefit generated to these groups. Thus it is difficult to conclude what exactly is the nature of SFDA programmes as an equity policy. However, on the basis of the above enumeration, here we would attempt to hypothize about the nature of the SFDA, as an equity policy.

For judging the nature of the SFDA programme as an equity policy, here the following three possibilities have been conceptualized.

Under an equity programme there could be a situation where most benefits from the equity policy go to the neediest and least benefits to the wealthiest sections of society, when such is the nature of the equity policy, it advances society towards a socialist goal and through these programmes, a society can become more equalitarian. This type of equity policy can be called a "substantive equity policy."

Second, an equity policy primarily helps the group living at the subsistence level. Here the equity programme primarily helps these groups to retain the subsistence level of living by preventing further deterioration of their living conditions. Thus it helps to maintain the equilibrium and prevent disruption of the existing social order. When the consequences of an equity programme are such, it does not reduce inequality in society. Here, an equity policy rather works as a mechanism to maintain the existing system of inequality but takes care of the residual problems. When an equity policy performs these functions it can be called a "subsistence equity policy."

There could be a third scenario under an equity policy where the equity policy neither provides most services to the neediest and least to the wealthiest nor does it work as maintainer of the subsistence level of living. Here, mainly because of political reasons, equity programmes are introduced on paper and they are mostly left unimplemented or misimplemented. This type of equity policy can be termed as a "symbolic equity policy."

In the light of the above categorization of equity policy, now we will try to ascertain the nature of the SFDA as an equity programme.

The SFDA programme has helped the poorest group of our society, namely, the agricultural labourers, to a considerable extent in the form of additional employment and increased wage. But it cannot be categorized as a "substantive equity policy", as this very process also has contributed to enhancing the income of the wealthier sections of the rural community, namely, better-off farmers, traders of agricultural inputs, etc.

At the same time the SFDA programme cannot be written off as mere symbolic equity policies for the very fact that it not only helped the wealthier sections of rural community, but also the poorest sections of society, the agricultural labourers. There is a mix of both the better-off and the worse-off sections of the rural community being benefited by the SFDA programme.

It appears that the SFDA programme as an equity policy has acquired characteristics more akin to a "subsistence equity policy." It has generated a process which, on the one hand, has benefited the better-off sections of the rural society and, on the other, the same process has taken care to benefit the worse-off sections of the rural community only to the extent that they do not degenerate to a below-subsistence level of living. Thus, the SFDA programme, instead of reducing the existing inequality, will maintain and probably add to the existing inequality, though not to the extent of disrupting the present conditions gravely. So long as the nature of the equity programme remains as such, it will not advance the socialist goal of our society, but will function only as a preventer of a major disruption of society due to increased inequality.

To sum up, the equity-oriented rural development in India originated more as a corrective measure than as an egalitarian commitment of the state policy. However, egalitarian considerations did play a role in influencing the policy recommendation. At the same time, the policy in the process of operationalization, under the influence of the prevailing administrative mechanism, property relations in the rural society and emerging agricultural technology deviated from equity-orientation. However, it appears that the prevailing market mechanism interacting with this situation resulted in a process whereby the very process (initiated by the SFDA) which benefited the well-off sections of the rural community also took care that the worse-off of the rural community are able to maintain a subsistence level of living. Thus, it appears that in Indian society the equity-oriented rural development is not operating as a "substantial equity policy" nor is it just a "symbolic equity policy." It appears to be more akin to a "subsistence equity policy." Therefore so long as the equity-oriented rural development in India operates in this fashion, it will not contribute to the advancement of the egalitarian goals of society. It will, on the contrary, further inequality though taking care to prevent a major disruption of the social order.

Notes

1. Following are some of the equity-oriented rural development programmes introduced during 1970s and 1980s:
   Small Farmers Development Agency Programme (Marginal Farmers and Agricultural Labourers Programme), pilot projects for Tribal Development, Tribal Area Development Programme, Drought Prone Area Programme, Hill Area Development programme, Desert Development Programme, Antyoday (Redemption of the last and the lowest) Programme of Rajasthan, Employment Guarantee Scheme of Maharashtra, Crash Scheme for Rural Employment, Pilot Intensive Rural Employment Programme, Food for Work Programme, Integrated Rural Development Programme, National Rural Employment Programme, etc.
2. Like SFDA, IRDP is a target group-oriented programme and both these programmes have the same approach of asset endowment by providing subsidy,

loan and other infrastructural facilities. Both the programmes share the same administrative arrangement where the district level SFDA project office has been replaced by district Rural Development Agency. However, unlike SFDA, IRDP's scope extends beyond agriculture and it covers all the blocks of the country. Thus IRDP can be considered as an extension of the SFDA programme.

3. Reserve Bank of India (1969) : Report of the all-India Rural Credit Review Committee, RBI, Bombay.
4. *ibid*, p. 537
5. *ibid*, p. 5
6. *ibid*, p. 536
7. *ibid*, p. 6
8. *ibid*, pp. 578-9
9. *ibid*, p. 537
10. *ibid*
11. *ibid*, p. 581
12. *ibid*, p. 537
13. Department of Agriculture, Govt. of India 'Central Sector Scheme for Marginal Farmers and Agricultural Labourers' Information Unit, Directorate of Extension, Ministry of Food, Agriculture, Community Development and Cooperation, New Delhi, 1971.
14. Benor, Daniel, et al., *Agricultural Extension-The Training and Visit Scheme,* — World Bank, Washington, 1977, p. 14.
15. Gaikwad, U.P. *Small Farmer: State Policy and Programme Implementation*, NICD Hyderabad, 1972, p. 2.
16. Swaminathan, M.S. 'Indian Agriculture at the Crossroads, Presidential Address, 37th All-India Agricultural Economics Conference, New Delhi, *Indian Society of Agricultural Economics*, 1977, pp. 18-29.
17. Ministry of Agriculture and Irrigation, Govt. of India, 'Report of the committee on Panchayati Raj Institution', Department of Rural Development, Ministry of Agriculture and Irrigation, New Delhi, para II, 1978, p. 32.
18. Mandal, K.S. *'Inequality and Development in India - A Study of an Equity-oriented Rural Development Programme'*, Unpublished Ph. D thesis submitted to the Indian Institute of Technology, Bombay, 1981, pp. 111-146.
19. National Commission of Agriculture, Govt. of India, 'Interim Report on Reorientation of Programmes of Small Farmers and Marginal Farmers and Agricultural Labourers Development Agency', Summary of Recommendation, New Delhi, 1973 p. (i).
20. Small Farmers Development Agency, Hooghly, 'Supplementary Project Report and Progress Report up to 1975-76', Hooghly, West Bengal, India, 1976, p. 6.
21. *Economic and Political Weekly,*' Institutional Change: A New Dimension', Vol. XIII, No. 8, 1978, p. 912.
22. Swamy, S. Dalip and Gulati, Ashok 'From Prosperity to Retrogression — Indian Cultivators During the 1970s, *Economic and Political Weekly,* vol. XXI, Nos. 25 and 26, June 21-28, 1986, A57-59.

tion and other infrastructural facilities. Both are programmes that are same administrative arrangement where the district level SFDA project earlier has been replaced by the District Rural Development Agency. However, unlike SFDA, IRDP's scope extends beyond agriculture and it covers all the blocks of the country. Thus IRDP can be considered as an extension of the SFDA programme.

3. Reserve Bank of India (1969), Report of the All India Rural Credit Review Committee, RBI, Bombay.
4. Ibid., p. 537
5. Ibid.
6. Ibid., p. 756
7. Ibid., p. 6
8. Ibid., pp. 753-4
9. Ibid., p. 177
10. Ibid.
11. Ibid., p. 538
12. Ibid., p. 539
13. Department of Agriculture, Govt. of India, Central Sector Scheme for Marginal Farmers and Agricultural Labourers: Integrated Unit, Directorate of Extension, Ministry of Food, Agriculture, Community Development and Cooperation, New Delhi, 1971.
14. Benor, Daniel et al, *Agricultural Extension: The Training and Visit System*, World Bank, Washington, 1977, p. 14.
15. Gaikwad, V.R. *Small Farmers: State Policy and Programme Implementation*, NICD, Hyderabad, 1971, p. 2.
16. Swaminathan, M.S. Indian Agriculture at the Crossroads, Presidential Address to the Indian Agricultural Economics Conference, New Delhi, *Indian Society of Agricultural Economics*, 1979, pp. 18-29.
17. Ministry of Agriculture and Irrigation, Govt. of India, Report of the Committee on Panchayati Raj Institutions, Department of Rural Development, Ministry of Agriculture and Irrigation, New Delhi, part II, 1978, p. 32.
18. Mandal, K.S. *Technology and Development in India: A Study of an Equity oriented Rural Development Programme*, Unpublished Ph.D thesis submitted to the Indian Institute of Technology, Bombay, 1981, pp. 111-146.
19. National Commission on Agriculture, Govt. of India, Interim Report on Reorientation of Programmes of Small Farmers and Marginal Farmers and Agricultural Labourers Development Agency, Summary of Recommendations, New Delhi, 1973, p. [illegible].
20. Small Farmers Development Agency, Hooghly, Supplementary Project Report and Progress Report up to 1975-76, Hooghly, West Bengal, India, 1976, p. 6.
21. [illegible] *Economic and Political Weekly*, Institutional Change: A New Dimension, Vol. XIII, Nov. 1978, p. [illegible].
22. Swamy, S. Dalip and Gulati, Ashok, From Prosperity to Retrogression—Indian Cultivators During the 1970s, *Economic and Political Weekly*, Vol. XXI, Nos. 25 and 26, June 21-28, 1986, pp. [illegible].

Index